JUGAAD
TIME

///

ANIMA Critical Race Studies Otherwise

A series edited by Mel Y. Chen and Jasbir K. Puar

JUGAAD
TIME

Ecologies of Everyday Hacking in India

/// AMIT S. RAI ///

Duke University Press   Durham and London   2019

Designed by Courtney Leigh Baker
Typeset in Warnock Pro and Scala Sans Pro by Copperline Book Services

Library of Congress Cataloging-in-Publication Data
Names: Rai, Amit, [date] author.
Title: Jugaad time : ecologies of everyday hacking in India / Amit S. Rai.
Description: Durham : Duke University Press, 2018. |
Series: ANIMA | Includes bibliographical references and index.
Identifiers: LCCN 2018027897 (print)
LCCN 2018037764 (ebook)
ISBN 9781478002543 (ebook)
ISBN 9781478001102 (hardcover : alk. paper)
ISBN 9781478001461 (pbk. : alk. paper)
Subjects: LCSH: Technological innovations—Social aspects—India. |
Economic development—Social aspects—India. | Digital media—Social
aspects—India. | Technology—Social aspects—India.
Classification: LCC HC440.T4 (ebook) | LCC HC440.T4 R34 2018 (print) |
DDC 338/.0640954—dc23
LC record available at https://lccn.loc.gov/2018027897

Cover art: The 6th Floor Collective, *untitled landscape from
the series "we're lovin' it,"* 2017. Acrylic on canvas, 100 × 160 cm.
Image courtesy of the artists.

To my mother,
Meera Mukherjee Rai,
1939–2015

Contents

Preface

A well-performed jugaad (hack) never fails to bring half-admiring, half-disapproving, and half-curious (it doesn't add up!) smiles to people's faces. Following what William James once said of fear and running, we smile before we admire. There is a certain intuition of the porosity of connectivity with the world that jugaad practice activates and makes ecological, even joyous. Today, perhaps uniquely in history, jugaad is a joyous passion. What is the time of that proleptic smile? It is, strictly speaking, the duration of a certain passage from affection to affection: jugaad's affective passage. Where are the spaces of and for jugaad practices? They operate within and against the plasticity—both neural and spatial—of India's "smart cities." Together, these space-times, plastic and durational, express the variable powers of emergent properties of nonlinear but feedbacked assemblages of affect, matter, policy, culture, biology, perception, value, force, sensibilities, practices, and discourses. Jugaad, as a practice of postcolonial practical reason and in its very timeliness, forms one way into and out of these assemblages.

This book emerged out of the changing political and personal landscapes shaping a set of collaborative researches into the politics of neoliberal technologies in India, the UK, and the US. In September 2010 I left a tenured position in an English department at a large state university in the US to take up a position as Lecturer in New Media and Communication at a then left-leaning business school in London. By then I had just finished a year-long research project on gender dynamics in India's fast-growing mobile phone ecologies, focusing on how to pose effectively questions of embodiment and perception in the contexts of cultures of pirated workarounds (jugaad) in digital technologies. Developing the concept of ecologies of sensation drawn from my study of new media in *Untimely Bollywood*, I was following an intuition that new political structures were emerging at the level of the body's habits, understood as always multiply situated processes of (dis)ability, control, and becoming, and that central to the story of this emergence was the then-nascent mobile phone ecology exploding differently throughout India, changing how people were relating to each other and to their so-called smart cities.

Jugaad Time constructs a heterodox interdisciplinary perspective to consider the potentialities and actualities of pirate digital cultures in India. This construction proceeds through a paratactical assembling of experimental diagrams of a commonly appreciated, but also suspect social practice, one that happens always and only at the volatile intersections of vectors of power and inequality. Jugaad is a Punjabi word that means workaround, hack, trick, or make do; today in ultranationalist, globalized India, it is both hailed and derided as a characteristically nationalist form of frugal innovation and also a possibly mediocritizing habit for shortcuts (see Mashelkar 2014; Radjou et al. 2012). The ethos of jugaad has become the dominant framework for economizing (or, more to the point, squeezing) labor power in neoliberal India, and it is pervasively used in emergent digital cultures. While contemporary neoliberal discourse has focused on jugaad as innovation, *Jugaad Time* seeks to develop a political philosophy of jugaad as an embodied ethics of becoming in India's caste- and gender-stratified smart/data cities.

This study begins with questions that are crucial to affective ethnography: in contemporary digital practices, what is the ontology of a jugaad event, and how can one practically understand its material effects and processual histories? This question leads, through a series of experimental parataxes, to what I will call a hacking empiricism. By empiricism I mean to refer back to the history of American pragmatism and British empiricism highlighted in Gilles Deleuze's work on the history of philosophy; in Deleuze's active creation of concepts buggered from this minor philosophy, a more nuanced understanding of perception in ecologies of sensation has begun to emerge. By hacking I understand a global but inchoate movement of workaround, informal (or "disorganized"—see Athique et al. 2018), extralegal, democratic, subaltern, collective repurposing of found materials shifting ecologies from relative stasis to absolute flux, at times and usually in the interests of narrow class segments, in innovative and "game-changing" ways. Together, hacking empiricism is a self-reflective practice of linking a problem—for instance, the problem of "why hack?"—to both its ecology of sense and sensation and to its processes and dimensions of change. Affective ethnography, as I will argue in the introduction, experiments with the critical capacities to affect and be affected by lines of flight and countergenealogies in and of India's City-with-No-Rights. Affective ethnography poses the transcendental conditions of possibility for radical and revolutionary action-potentials meshed in ecologies of workaround practices. Clearly, through this paratactic network of rhizomes, the works of Gilles Deleuze and Félix Guattari will also form crucial nodes in this construction. Their work on affect and ecology is mobilized in an affirmation of jugaad as the expression of a subaltern and autonomous sensibility besieged, maimed, imprisoned, and controlled by mutually ramifying and contradictory (it doesn't add up!) forms of domination, exploitation, dispossession, commodification, monopoly, and habituation.

Guattari, of course, helped develop a diagrammatic praxis in different domains, from psychology to political organizing. This diagrammatic method, which forms the space-times of what have come to be called affective ethnographies, affirms both the creativity of revolutionary movements throughout history and the struggles involved in their becomings. *Jugaad Time* paratactically assembles a nonrepresentational diagram of the social, economic, political, and ecological vectors (tendencies, capacities, functions) of jugaad by considering the various sensorial, algorithmic, material, and territorial makeshift infrastructures enabling digital cultures in India today. This is a key focus of *Jugaad Time*: developing a pragmatic ecology of jugaad. In the context of postcolonial India, in terms of both makeshift capitalist and noncapitalist infrastructures for different supply-chain strategies as well as an implicit pirate ethics of commoning resources against neoliberal dispossession, this method develops a political diagram of patterned but unpredictable becomings within and against ecologies of hacking practices.

This diagrammatic method draws out the functional and potential resonances traversing the health, ecological, and labor conditions for workers and communities in different nodes of algorithmic capital. Hacking and informalized workaround practices of increasingly precarious workers have a material existence in logistical and supply-chain processes that traverse and exceed capital. For instance, the research and development of mobile devices and value-added services, tied to new algorithms and sensibilities of forwarding, while it has provided vast riches to the mobile value-added services industry (a creative industry if there ever was one), have as a condition of possibility the violent and unsustainable coltan mines in the Congo and Brazil. This transcendental empiricism of the value-added affection would also pass through the mobile assembly plants encircled by suicide nets, black-market distribution networks, informal points of sale, smart city/Big Data sensors, back through the new technological innovations in political marketing in, for example, Prime Minister Narendra Modi's digital election campaign machinery, the rise of rightwing Rahstriya Swayamsevak Sangh propagandizing through social media on US and UK university campuses, the Islamophobic co-branding of Israel and India, and so forth.

Considering this far-reaching media ecology as both volatile and value generating, *Jugaad Time* develops a method that aims to take empiricism in the direction of pragmatic experimentations in becoming Other(s). By taking as its primary examples everyday practices for the marketing and hacking of digital cultures, *Jugaad Time* engages contemporary media studies and anthropological and practice-based methodologies, and it affirms the different political projects of queer Marxism, Dalit emancipation, postcolonial workerism, and posthumanist feminism.

Diagrammatic Methodology and Its Sociohistorical Context

As I will elaborate more fully in the introduction, as method, *Jugaad Time* makes the case for an ecological encounter with social practice that is itself best understood as a pragmatic, ad hoc, networked approach to an obstacle. Potential and actual at once, the ecology and social history of the specific pragmatism employed in the practice of jugaad is a mode of sufficient reason proceeding through intuition as much as probability (Ansell-Pearson 2001; Bergson 1988; Deleuze 1988a). More concretely, I have been part of a collective research practice involving colleagues at the Tata Institute of Social Sciences' Centre of Media and Cultural Studies, as well as with the following research assistants: Anisha Saigal (Delhi), Shiva Thorat (Mumbai), and Rachna Kumar (Mumbai). Through mostly what is called "snowball" sampling, we were able to interview and codiagram with different jugaadus in Bhopal, Mumbai, Bangalore, and Delhi. These actants, some of whom shared our left politics of anticapitalist, feminist, revolutionary/radical, and emancipatory critique, many of whom didn't or were indifferent, became participants in a collective experiment in pragmatizing in mobile ecologies: what workaround repurposes our relation to technology and technique itself?[1]

As a term, jugaad has a wide range of colloquial uses throughout contemporary neoliberal India, and it has been thematized explicitly in South Asian media for decades. In that sense we can understand it as always already doubled: as the idea of an idea (or what Spinoza called method), the sensory motor circuit of jugaad has an intensive and highly mediatized history. More, it is precisely this feedbacked and intensive quality to jugaad that allows for a renewed engagement with a heterodox political economy of contemporary digital capitalist control and its parasitical and autonomous pirate kingdoms (Sundaram 2009). This includes situating a transnational capitalist class of South Asians more and more integrated with the neoliberalizing agenda of the postcolonial state, a state prone to personality cults, family dynasties, trustee capitalism, corrupt bureaucracy, ultranationalisms (and other postcolonial *ressentiments*), and one that is increasingly invested in global oligopolies based around several core logistical integrators (e.g., Coke, Amazon, or Disney). This political struggle to manage contemporary forms of neoliberal exploitation, capitalist crisis, and social control necessarily produces national, regional, and cultural forms of legitimation and struggle; indeed, jugaad as the figure of the makeshift assemblage of frugal innovation has been unevenly integrated into this multiplicity of processes. *Jugaad Time*'s heterodox political economy of digital control affirms the vibrant ecologies of thousands of pirate kingdoms (Larkin 2008; Sundaram 2009), some of which, some of the time, mutate decisively from events that hack through and queer (or unpredictably and immeasurably intensify) the probable distributions of hypermodernity.

Michel Foucault wrote that knowledge was not made for understanding, it was made for cutting: *Jugaad Time* is written in the spirit of developing pragmatic assemblages that jugaad different ways to exit both capital and the forms of subjectivity that are within and against it. Through the exploration of the everyday potentialities that haunt the habits, events, time-spaces, encounters, sensations, processes, infrastructures, perceptions, and entrepreneurial capture of digital cultures and their hacking in India today, *Jugaad Time* elaborates India's new abstract diagram: what is the force, sense, and value of the habituation of jugaad, and what specifically would effect its actual and potential collective reorganization? Indeed, what has done so historically? Analyzing jugaad at once as an event in the molecular and molar histories of material processes, and as the name of a contemporary practice of capitalist value creation (known in business and management through the term "frugal innovation"), *Jugaad Time* proposes a new political philosophy of an embodied machinism that variably participates in numerous animal, bacterial, chemical, technoperceptual, and political becomings. *Jugaad Time* thus takes seriously and yet quite lightly Gilles Deleuze's affirmation of becoming in the potentializing passage between affective states: (de)habituation is nothing other than this negotiated potentialization (see Bhabha 1994; Deleuze 1988b). The specific postcolonial time of jugaad becomes relevant to critical practice as repurposed bodily capacities potentialize affective states on a plane that we can better diagram through connecting relations and functions of force, value, and sense in contemporary digital media infrastructures (both pirated and corporate). In the pages that follow, I develop an understanding of the media assemblages of jugaad, how jugaad knowledge—subjugated and subalternized—proliferates as a countermemory of domination, privatization, and dispossession, even as it gives practical form and revolutionary sense to capitalist value-creation strategies. I begin with an engagement with different practical sites for the emergence and distribution of jugaad media, such as mobile phone memory-card movies, video compact discs, peer-to-peer file sharing, and so on, thus linking together resistant forms of negotiating the digital irreducible to capitalist strategies of value capture. I hope to show that what jugaad affirms is the necessity of a political strategy of exiting the relations of value, sense, and force demarcating both the "moral universe" of capitalist oligopolies globally, and the forms of subjectivity that have developed to critique those systems of reference.

A jugaad can be extralegal, but it is always multiply within (as embodied subjectivity and relation of production) and usually implicitly against capital (see Berardi 2008; Mandarini 2005; Tronti 2005). In other words, in a jugaad event the boundaries of what is both possible and necessary become plastic through a more or less pragmatic experimentation in habits, capacities, material processes, collective enunciations, and assemblages. The jugaad event presents us with a ready interpretation, and its criticism would show it to be easily adapted to the interests of capital: jugaad would then function as the product (as a fetishized method of frugal innovation) that covers over processes of subalternization, precarity, depoliticization, and dispossession intensifying throughout India. The wager of *Jugaad Time* is that these processes entail both an actual and potential politics.

In other words, queer, feminist jugaads body forth other diagrams for collective practice that effectively differentiate them from a neoliberal jugaad. Thus, as a radical practice, the event of jugaad invites potentializing questions to be posed of what difference queer and feminist assemblages body forth in this force field of potential. Many of today's struggles for democratic, nonhierarchical, and free social space in India work within and against these relations of force, jugaading the future into the present; for instance, as we shall see in chapter 3, in the Why Loiter? movement in Mumbai, initiated by three feminists, Shilpa Ranade, Samira Khan, and Shilpa Phadke, a dispersed collective intervenes practically in transforming political, economic, social, gender, sexual, and kinship relations in India, adopting social network and mobile technology strategies for both general activism and more specific kinds of hacktivism. Why Loiter? poses its question by foregrounding another: whose right to the city?

The ecology of sense and sensation that jugaad events operate through, and happen in, requires specific pirate infrastructures. Thus, in the diagrammatic method of jugaad we proceed first by counteractualizing from product-event to potential process-infrastructure. Again, this moves the method of *Jugaad Time* beyond mere critique, argumentation, explanation, correcting, or demonstration toward an urgent and practical reconsideration of the resources, capacities, and affordances that are necessary to hack into and against contemporary capitalist media ecologies.

The political project of *Jugaad Time* makes an affirmation of becoming through a counteractualization of the jugaad infrastructures of postdigital cultures. This political project in other words is nothing other than the method on another intersecting plane of value, sense, and force. The measured quanta of this value may already have a very warm home as frugal capitalist innovation, but it references other systems of sense, value, and force shaped by the histories of insurgency, maroonage, mutiny, refusal, the commons, and exit that brought new relevance and new understanding to subaltern agency and the autonomy of the oppressed. Importantly, contemporary media studies and cultural anthropology of information industries (with a renewed focus on rapidly "developing" India) have put new media cultural practices in relation to globalization, mediascapes, embodiment, political and human geography, assemblages, government policy, digital cool, colonial administration, infrastructures, and ultranationalism as an "affective excess marked by a hyperperformative jingoism" (Liang 2009; see also Ajana 2013; Anand 2011; Aneesh 2006; Ash 2010;

Birkinshaw 2016; Castells 2015; Dasgupta and Dasgupta 2018; Easterling 2014; Gabrys 2014; Mankekar 2015). The diagramming method followed in *Jugaad Time* traverses these important analyses to pose nonrepresentational and materialist questions of different forms of media power. Thus, I offer up a pragmatic ethnography of a mobile value-added company in the Delhi national capital region through strategic narratives of acts of consumption that become potential sources of new productivity, as Mobile Value Added Service management information systems (MIS) reports correlate fresh data with patterns of use in other data sets. This notion of an ecological and process-oriented historical materialism affirms forms of exiting practices that never cease leaving the court/clinic of commentary and interpretation by thinking through assemblages of action and acting through assemblages of thought. In that sense, *Jugaad Time* is an extended and sustained reflection on collective practices of habituation, informatization, and counteractualization in pirate economies in India.

Given that it is one of the most competitive and fastest-growing mobile markets in the world, India's heterogeneous digital culture bodies forth subaltern diagrams of mobile hacking, neoliberal consumerism, digital control, media piracy, embodied perception, technological habituation, and new media assemblages. This method, as I have suggested, has been guided by the question of how to understand effectively the knots of these technoperceptual relations, their historical emergence, and their various phase transitions. Transdisciplinizing a wide range of research, *Jugaad Time* moves beyond a social constructivist digital culture where the psychologized digital is never not a scene of representation and the abyssal meaning of bad consciousness, arguing for a pragmatic approach to digital ecologies and their hacking. Far from a mediocritizing workaround, the social practice of jugaad shows the urgency of creating new micropolitics for commoning private property while working around capital and its regimes of control (e.g., peer-to-peer networks). *Jugaad Time* then presents an untimely practice of diagramming and queering relations of sense, force, and value in technoperceptual assemblages, such as the location-based data flows of mobile phones.

By taking the affective capacities (the bodily capacity to affect and be affected) as the starting point for understanding the history, mutations, effects, force, value, and sense of digital cultures in India, *Jugaad Time* situates the mediatization of the practice literally at the threshold of a new hypermodernity, understood as both controlled modulation of capitalist crisis and the counteractualizing effects of these at-times untimely and queer practices. Throughout the study, I argue for a methodology of affect-as-capacity, which shifts the focus of attention away from language, discourse, and representation toward habituated and emergent sensations in historically specific media assemblages. Thus, in terms of method it brings together a pragmatic empiricism (drawing on the work of the new materialists in social geography, queer studies, and postcolonial studies) with wide-ranging historical diagrams of digital cultures in Mumbai and Delhi, affirming, with Patricia Clough, the "feelings, vibrations, rhythms and oscillations coming up from the streets" (2010, 230). These affective diagrams are coupled with broad archival research on the transformation of telecommunications in colonial and postcolonial South Asia, as well as (non)narrative speculations on the future of potential and the potential futures in/of these ecologies.

Jugaad Time elaborates a concept developed in conversations with Patricia Clough and Jasbir Puar, and in an earlier work, *Untimely Bollywood: Globalization and India's New Media Assemblage*, namely, ecologies of sensation. Drawing on Guattari's definition of assemblage as a prepersonal practice, or a kind of style, a creative mutation that binds an individual or a group consciously or unconsciously, *Jugaad Time* shows how technoperceptual habituation in India feeds back into historically variable and also dynamically open ecologies of sensation. Definite but fuzzy ecologies of sensation enable practices of jugaad that make perceptible relations of motion (assemblages) between a body's neurological capacities activated in habituated gestures, perceptions, intensive flows, associations, affect-time-images, sensations, and the distributed kinesis of the postsovereign subject through nonlinear space-time. The changing emergent capacities of digital-human assemblages (distributed networks, evolutionary algorithms, crowdsourcing in social networks, tracking, monitoring, recommendation algorithms, datamining, peer-to-peer file sharing, multimedia platforms, free and open-source computing, value-added marketing strategies, and always-on connectivity) have become a hotly contested domain of struggle helping us to re-pose the question of the (de)habituation of different populations through networked hacking practices. In *Jugaad Time*, I engage a broad range of digital communication practices to show how specific forms of sexuality emerge from historically variable but always-potential ecologies of sensation, as new forms of digital memory and habitual embodiment mesh together to produce new forms of value, sense, and force. To work around an obstacle is in some way to reenchant the world through a canny giddiness in the face of its infinite sponginess: working through an intuition animating the method of hacking empiricism involuting of gender, sexual, caste, class, raced, and embodied effects. Thus, *Jugaad Time* takes up and repurposes contemporary academic discourses on the feminist cyborg, antiracist futurism, the politics of affect, autonomous biopolitics, hacking digital media assemblages, the end(s) of social constructivism, and queer assemblage theory. The time of jugaad is the time of sexed and gendered bodies, smart cities, and analog and digital technologies in flux, in and out of tune/resonance, open to their outsides, and increasingly and complexly within and against the logistical machinery of racialized and misogynist capital. Any political ecosophy commensurate with its overthrow will need the methodological panache of the jugaadu.

Acknowledgments

A book of this kind is already quite a crowd: collaborating researchers, jugaadu participants, and all the nonhuman agents whose voices are expressed in formative processes. In terms of humans with names, I would like to thank Shilpa Phadke; Smita Rajan; B. K. Rai; Shail, Sudeep, and Divya Shrivastava; Abhay Sardesai; Sarah Elise Fryett; Stefano Harney; Gerry Hanlon; Ranjit Kandalgaonkar; Vinita Gatne; Tarek Salhany; Polly Phipps-Holland; Nagesh Babu K. V.; P. Niranjana; Anjali Montiero; K. P. Jayasankar; Faiz Ullah; Nikhil Titus; Priya Jha, Gerry Hanlon; Matteo Mandarini; Sadhvi Dar, Yasmin Ibrahim; Hypatia Valourmis; Shiva Thorat; Anisha Saigal; Rachna Ramesh Kumar; Renu Savant; Matteo Mandarini; Liam Campling; Elena Baglioni; Ashwin Devasundaram; Jenny Murphy; Keiko Higashi; Bibi Frances; Gini Simpson; Lois Keiden; Arianna Bove; Erik Empson; Jasbir Puar; Patricia Clough; and Shambhu and Leena Rai. I have been supported by the Fulbright scholars program; QMUL internal grants, the AHRC, and in another but no less material way by the radical collectives at the Common House, Bethnal Green.

A POLITICAL ECOLOGY OF JUGAAD

Jugaad and Ecology

Jugaad Time: Ecologies of Everyday Hacking in India emerged out of partisan research into digital media in postliberalization northern and western India. I say "partisan" in the sense of politicizing: in the global North, neoliberal capital as a social and economic formation attempts to exclude "the people," "the masses," or "the multitudes" from political participation in order to separate everyday life in the so-called free marketplace from social and economic emancipation. In postcolonial India, neoliberalism has "progressed" in the contexts of specific caste and class politics, urban/rural divides, state-centered versus autonomous feminisms and workers' movements, varieties of religious and nationalist chauvinisms, and the emergence of a distinctive and vibrant queer politics.[1] My attempt in this book has been to situate in a postcolonial frame a certain vector of becoming associated in Western criticism with the tinkerer, who in this study takes the form of a preindividually primed and collective subject, within and against the postcolonial cunning of neoliberal reason and embodied in pirated digital social practices. These practices emerge in contested political ecologies themselves within and against capital, authoritarianism, and static/state identities. My research trajectory was shaped by a year-long research fellowship in Mumbai and Delhi, during which I began to focus specifically on questions of gender, political ecology, heterodox economics, and mobile phone hacking. Gradually, through research conceptualized and conducted in collaboration with colleagues and students at the School of Media and Cultural Studies, Tata Institute of Social Sciences (Chembur, Mumbai), I began to develop practical genealogies or, what I will call in this book, diagrams of people's media practices in India's huge "informal" sector (around 90 percent of the economy). This sector, and especially in its relation to the emergence of what Athique and colleagues (2018) have called India's media economy, can be understood to have interrelated ecologies of dispersed and noncontractual logistics and organization (commonly referred to as the informal or disorganized economy—see Rajadhyaksha in Athique et al. 2018; Venkatraman 2017). It was in the course of this research, while analyzing corporate Indian media strategy, that I first encountered the Hindi/Punjabi colloquialism "jugaad."

That word, repeatedly featured in people's self-presentation of their meshed media practices and work-related strategies in everyday life, is a reference to a sometimes elegant, but always makeshift way of getting around obstacles. Jugaad practice abducts forces to yield a new arrangement or assemblage; indeed, the minimal unit of any jugaad practice is the assemblage (Rai 2009). Consider, as a first approximation, assemblages distributed across two news reports from the *Times of India* (Mumbai edition), both dated March 24, 2010. The first shocked readers throughout the city: a twelve-year-old victim of serial sexual abuse had come forward to name her assailants. One of the main accused had recorded an MMS (multimedia messaging service) digital video on his mobile phone in which he and another friend raped the twelve-year-old girl. They used the threat of going viral with the MMS to coerce the girl into having sex with them (this use of the mobile's audiovisual function for violence against women has become commonplace in India's digital misogynist rape culture). The second report relates events surrounding another coercive use of the mobile, this time through SMS (short message service). The police detained five suspects, including one accused in a murder case, in Thane (a northern suburb of Mumbai) in order to probe their role in sending threatening SMS texts to the leader of the opposition party, Eknath Khadse (a Bharatiya Janata Party [BJP] member of the Maharashtra Legislative Assembly) and to the Shiv Sena leader Eknath Shinde. According to the report, after initial investigations of mobile phone logs, police found that the person who sent the threatening SMS texts had also telephoned the office of the Thane police commissioner around 1:30 a.m. These are only two disturbing and sensational stories of India's mobile phone ecology. What are the conditions of their possibility? One sense of jugaad that I will return to in my conclusion resonates with misogynist uses of new communication technologies to shame, coerce, torture, and troll women into a generalized and enforced silence. This violent silence central to misogynist lifeworlds has been systematically and effectively addressed in feminist organizing in India, seeking to disrupt its functioning and overthrow the sources of its genesis. Jugaad in this scenario becomes a way to quite literally hack a woman's life. The circulation of these force-images in print, on the internet, on satellite TV, on a mobile phone, or through rumor and word of mouth produces various kinds of sensations, from the violence of sexual assault, or the pornographic titillation of masculinist sadism, to the concern over the sexual risks associated with proliferating communication interfaces. At stake in both stories is the practical use of mobile phones in India today: as weapon, lever of value and force, convergent technology, surveillance device, superpanoptic gaze, and viral sense machine. The mobile, taking over the role of the webcam from the 1990s, has become the instrument of choice in what Paul Virilio presciently called "generalized snooping" (Virilio 2005). Indian populations across caste, class, religion, gender, sexuality, region, and generation vectors are undergoing a rapid and expanding rehabituation in this jugaad of mobile communications. Rehabituation in jugaad time enfolds the multiplicity of different histories, timescales, intensities, durations, potentialities, actualities, vectors of power and performances of identity. A new form of life is emerging in postcapitalist India, and this biopolitical morphogenesis is emerging from a newly potentialized ecology of sensation. In what sense is preindividual, precognitive sensation being potentialized and revalued in India through the mobile phone? In what follows I give specific examples from my field work in Mumbai, New Delhi, Noida (Uttar Pradesh), Seore, and Bhopal (Madhya Pradesh) to indicate some emergent nodes of a speculative diagram through which an experimentation in the force, sense, and value of mobile ecologies of sensation can proceed.

Another parataxis emerges in an interview with a mobile value-added services (MVAS) project manager in North India, excited about the "scope of VAS in India." As a project head in his company, he was very excited about the innovations and possibilities for such services and about the scope of mobile telephony more generally. He went quickly through some highlights of the mobile "jugaad revolution" in India: in 2003–2004 Reliance mobile revolutionized the industry by bringing costs down to about Rs. 1,200/month; today, 95 percent of customers are prepaid cardholders. The pattern of communication for most people is to receive calls, which is free in India; most people do not own a handset for more than twelve months. The aim of VAS is perceived value—what can be termed self-expression products: callback tunes, wallpapers, mobile accessories, and the like. I wrote afterward, "The vivid sense I got from him was that things are changing rapidly; there is a lot of innovation happening in the industry, and the regulations are coming down. Lots of scope."

The specific problems in delivering MVAS are rooted in the processes that are covered over by this industry-scale, financializable view of adding value. A project manager at another MVAS company in North India is working on developing mobile-ready services for Kriluxe Paints. In the paint cans is a specially covered scratch card through which people can avail themselves of various prizes. One of which is a 50-rupee credit for your prepaid card, or Rs. 10 off your postpaid card. The program will give out other big prizes, a TV and a scooter, as well. Now the problem has been to get credit from different mobile vendors all over India because today the system is malfunctioning due to a virus (not uncommon). Basically the project manager and her team (the entire staff is organized into project-specific teams) have had to enter this information by hand, even though they guaranteed that it would be automated. But overall it has been a success, and even before the official launch of the media blitz a buzz has started so much so that Kriluxe has been able to increase its market share nearly twofold. Such problems, and their financial reverberations, are endemic throughout the MVAS industry, and so it comes as no surprise that the dominant and yet minor practice of tweaking a given service till it works is called jugaad.

Such examples push us to think of jugaad a bit more creatively. We could start with the notions of abduction and experimentation: how do jugaad practices experiment with abduction? C. S. Peirce's notion of abduction can be described as a temporal process of tacking back and forth between futures, pasts, and presents, framing the life yet to come and the life that precedes the present as the unavoidable template for producing the future. For Vincanne Adams and colleagues,

Abduction names a mode of temporal politics, of moving in and mobilizing time, turning the ever-moving horizon of the future into that which determines the present. Abduction is a means of determining courses of action in the face of ongoing contingency and ambiguity. . . . Ideas about how to "move forward" are generated by tacking back and forth between nitty-gritty specificities of available empirical information and more abstract ways of thinking about them. In anticipation, abduction also acquires a temporal form: the tacking back and forth between the past, present and future. Abduction moves reasoning temporally from data gathered about the past to simulations or probabilistic anticipations of the future that in turn demand action in the present. Abduction thrives in the vibrations between the is and the ought, consummately modern yet augmented by anticipation in ways that undermine the certainties on which modernity thrives. (2009, 252)

Following on, Thrift notes that it is in fact "abduction or theory construction which is the outstanding characteristic of human intelligence. Abduction is the leap of faith from data to the theory that explains it, just like the leap of imagination from observed behaviour to others' intentions. While most explicit theories or abductions are wrong, our implicit ones about interactional others are mostly good enough for current purposes" (Levinson 1995, 254, qtd. in Thrift 2005, 466). It is this "mostly good enough" theory of abduction that best characterizes the pragmatism of jugaad. Pragmatically the aim in the scenario above is to create a brand-equity buzz among mobile phone subscribers and paint buyers through the perception that Kriluxe can give you good paint and the added excitement/value of a mobile phone–enabled prize drawing. This buzz tacks between the present, the past, and the future in the sense that it draws on current brand equity, associates that with a now-wow gimmick, and projects a future-oriented hype for the brand. This abductive jugaad keeps in play the full semantic range of abduction: guessing and also kidnapping. Despite regional differences, linguistic barriers, and computer viruses, the service will go on by any means, even if that entails the digital-becoming-analogue.[2]

Thus, jugaad figures forth a supposedly essentially Indian, but now suddenly revitalized style of activating the empiricist idea that "things do not begin to live except in the middle. In this respect what is it that the empiricists found, not in their heads, but in the world, which is like a vital discovery, a certainty of life which, if one really adheres to it, changes one's way of life? It is not the question 'Does the intelligible come from the sensible?' but a quite different question, that of relations. *Relations are external to their terms*" (Deleuze and Parnet 2007, 54–55, emphasis in original; see also Culp 2016, 42).

Jugaad is an everyday practice that potentializes relations that are external to their terms, opening different domains of action and power to experimentation, sometimes resulting in an easily valorized workaround, sometimes producing space-times that momentarily exit from the debilitating regimes of universal capital (Culp 2016, 17; Deleuze and Guattari 1994). In popular usage jugaad can refer to a savings account and its attendant ideology of insurance, the extralegal workaround practices of "informalized" or "disorganized" sector workers, the questionably legal workarounds and patronage deals that the local and central state apparatuses depend on, a pedagogy of empowerment within and between subaltern groups, a way of minimizing the risks of the future, and a way of working around domestic employer expectations by, among other things, digitally curating recipes. As we shall see, its practice can create metabolic imbroglios in ecologies of social reproduction as well.

Jugaad gradually became for me a way of posing better questions regarding media, neoliberalism, and politics in India by tracing relations external to their terms. These terms, which, as Muriel Combes in her forceful elaboration of Gilbert Simondon's transductive ontology shows, are relations of relations (2013, 17), focused attention on questions of value, and performances of gender and sexuality in mobile phone ecologies, and as my own method became ecological, the relations of power and affect within changing gender norms and sexual performativity questioned the fundamental dynamics of capital and accumulation under a Hindutva-driven neoliberalism. Thus, in India today these relations are undergoing several nonlinear phase transitions in terms of habits, and new political forms and social formations are emerging between the paving stones demarcated by the state in its authoritarian, quasi-entrepreneurial march to Hindutva capitalism.

This was in 2009–2010, when I was doing research at an "up and coming" MVAS company in the Delhi-National Capitol Region. It is difficult to describe that dizzying time without getting caught up again in the massive whirl of globalized media events and the often brutal and sometimes obscured forces that were at that time redefining a neoliberal "India shining." In these same cities that I was now conducting interviews, developing political and research contacts, and participating in old and new media cultures, the processes of masculinist and upper-caste Hindu chauvinisms, elite neoliberal globalization, speculative gentrification, corporate marketing and branding, enforced austerity, environmental degradation, and social and religious segregation had for many years been undermining India's democracy, and ironically, given the resurgence in various kinds of chauvinisms (religious, regional, linguistic, etc.), its local cultures.

But if, as Jacques Derrida once wrote, the future is an absolute monstrosity, and intuiting its patterned but unpredictable forces requires an ecological and collective practice of politicizing our meshed and intuited ontologies (Derrida 2016; Guattari 1995), practical research develops untimely transvaluations of all values (Nietzsche 1966). One mode of this transvaluation will proceed through a "hacking empiricism" that takes as its target and instrument the mystifying dialectic of what Alfred Sohn-Rethel recognized as central to the social synthesis of capitalist exploitation: mental and manual labor (Sohn-Rethel 1978, 4–6). This dichotomy is directly addressed and displaced in the social practice of jugaad. This book will attempt to diagram jugaad, a method that will be defined through a continually intensive and critically recursive paratactical diagramming, or affectively relating the virtual and actual ecology of its various resonances (and, more broadly, following Marx and others, its specific but open web of social and material relations). As I will argue in the pages ahead, the intensification and deterritorialization of capitalist flows of surplus value and profit in neoliberal India help us to situate jugaad's ecology of sensation.

JUGAAD IS THE PRECARIOUS logistical practice of "choice" in India's vast informal, disorganized, pirate, extralegal economy. As Athique and colleagues suggest,

> All of these large scale endeavours, public and private, can be collectively referred to as the "organised" sector. They are media structures constructed through acts of policy, regimes of regulation and the monopolisation of bandwidth in one form or another. Given the imperatives of import substitution as the guiding principle of the planning process, organised sectors such as broadcasting and electronics were

considered in terms of productive capacity. Large-scale publishing and the press were somewhat different, in terms of their private ownership and constitutional relationship to the democratic project. Nonetheless, market forces were rarely a driver of media development throughout the organised sector during the mixed economy era. Whilst indisputably important, the centralised and heavily regulated development of the planning era was nonetheless only one part of the story. Much of the development of India's media economy actually took place in an entirely different domain. By this, we are referring to India's informal economy, where a vast field of commercial activity takes place largely beyond the regulatory framework, and often beyond the purview of governance. A strict definition between the organised and unorganised sectors can be attempted on the basis of the number of employees or the existence of formal contractual relationships, but the essence of what Vaidyanathan calls "India Uninc." is probably too complex to capture by any simple measure. But, in general terms, the unorganised sector represents the aggregation of small-scale enterprises, non-contractual labour, reciprocal obligations and cash payments. (Athique et al. 2018, 10–11)

Jugaad is the "ethical" know-how of this (dis)organization in the spatial contexts of what Ananya Roy has termed the insurgent city; writing of the urban planning context of West Bengal, she notes the peculiar complicities of the jugaadu:

The fierce and bloody struggles in Nandigram seem to mark a break with such patterns of political dependence. And yet, they can also be understood as yet another instance of populist patronage, one where insurgent peasants are now bound to the electoral calculus of oppositional politics and the protection of the Trinamul Congress. Such forms of insurgence then do not and often cannot call into question the urban status quo; they can imagine but cannot implement the just city. And most of all, they depend on, and simultaneously perpetuate, the systems of deregulation and unmapping that constitute the idiom of planning. This is the informal city, and it is also an insurgent city, but it is not necessarily a just city. It is a city where access to resources is acquired through various associational forms but where these associations also require obedience, tribute, contribution and can thus be a "claustrophobic game." (Roy 2009, 85; see also Rawat 2015; Rawat and Satyanarayana 2016; Roy 2001, 2005)

FIGURE I.1. Municipal jugaad in Bangalore (2015).

Jugaad practice often works through or in parallel with this claustropho-bic game of associational patronage—some jugaads require a "source" (i.e., someone in power) to be tapped before the jugaad can even take shape. A jugaadu is also the neoliberal "confidence man." The work on the many different unintended, pirate modernities coevolving in the subaltern and Dalit smart city has highlighted the increasingly unequal, informalized, and informatized modalities of this neoliberal assemblage.

This book attempts to extend and broaden that analysis in the direc-tion of the specific affective relations and habituations that have come to dominate the ideologies of the smart city. I wish to show, in other words, that if understood ecologically, and as an ecology in itself, jugaad poses ontological and durational questions for social practice, urban space, po-litical agency, and embodied habit. To be clear, however, jugaad is not pre-sented here as any kind of political program for revolutionary practice or becoming. Indeed, in some sense the revolution will be anti-jugaad. Rather, jugaad is considered from the perspective of duration; one of these durations can indeed be understood as involved in a revolutionary becom-ing. In terms of habituation and logistics, jugaad experiments.

In this introduction, I first suggest a working definition of jugaad. Through a recent "nudge"[3] marketing campaign for the digital media organization Sulekha, we will come to situate jugaad practices in a surprisingly volatile social and political field of struggle, one in which the intercalated forces of gender, sexuality, class, religion, and digital technologies in India shift frames of reference in hacking ecologies, and in their practices revalue capitalist systems of sense, sensation, and power. I then try to further explore different ecological dimensions of this practice and its "affective image" in a consideration of a short documentary of a video pirate in Mumbai, *Videokaaran* (2012). I then draw the methodological implications of these sets of analysis and consider their resonance with contemporary postcolonial ecology studies. I conclude by situating the chapters to come in terms of what I consider the "diagrammatic" method of this affective ethnography of jugaad.

In jugaad practices, more or less sustainable ecologies are themselves transvaluated so that their functional connections, synchronicities, and embedded processes become objects of distributed human and nonhuman intervention. Sometimes a new value emerges from this intervention, as when peer-to-peer file sharing and hacker subcultures affirm the once and future freedom of the internet, but often not so much, as usually a jugaad guarantees at least for the short term, more of the same, as when a family in Delhi decides to hire private contractors to establish a direct water or electricity connection because either they have no current access or inadequate access. These kinds of jugaad rarely come out of any radical ecological consideration of the material effects of one's actions, and they are rather geared to money/time saving and/or productivity-obsessed, short-term self/family-enrichment (Culp 2016). Here, recalling Roy's argument, we mark again a certain danger in the practice of jugaad that will form an ongoing lever to consider the distinction between manual and intellectual labor that jugaad consistently overturns, if only then to generate a new value-added innovation (time savings, speculative currency, reputational rents). While a jugaad is usually shrouded in the mystifying discourse of individual creativity, that is, a labor of intellectualized imagination, as a collective if distributed practice of everyday life, it remains a pragmatic approach to intervening effectively in a volatile and increasingly precarious field of possibilities/probabilities, and unknown, but experimented with, forces, capacities, virtuosities, obstacles, bottlenecks, flows, and connections. An intensive paratactical approach to this subaltern and now increasingly hipster practice proceeds without prejudging its ethics and its politics: this diagrammatics will have dispensed with judgment. Instead, in an affirmative spirit of critical solidarity, I tease out the relations of relations in jugaad's varied becomings, an affirmation of an indomitable, non- and postcapitalist, but complicit creativity distributed and emergent in the millions of pirate kingdoms that have expanded and crisscrossed the neoliberal world, each immense and immeasurable, but susceptible to control (Hardt and Negri 1999).[4]

This then is the general purview of this study. I focus on the interwoven cultures of jugaad and digital media in India's smart cities. Through the analysis I attempt to link jugaad or hacking ecologies to changing patterns of media prosumption (consumption as productive), gender relations, and class/caste and religious politics. Throughout, I draw on interviews conducted by myself and research assistants (Ajinkya Shenava, Shiva Thorat, Rachna Kumar, and Anisha Saigal) between 2009 and 2017. These interviews were conducted through a kind of snowball sampling, and they were recorded, translated, transcribed, and anonymized. The aim of the interviews was to engage all participants in an open-ended exploration, linking questions of lived gender, caste, and class relations and technology and digital infrastructure. I offer the following as nodes in a critical diagram of neoliberal media ecologies, parataxes relating this relatively old, but newly mediatized practice and discourses of jugaad.

(Anti-)Jugaad and the Innovation Image

The field of struggle in which jugaad ecologies thrive emerges fragmented in contemporary Indian media representations. The overwhelming uptake of neoliberal values across India's media ecologies is one way to consider recent campaigns affirming an "anti-jugaad." Take, for instance, a recent TV advertisement and consumer behavior "nudge" campaign created by Ogilvy and Mather (India) for Sulekha.com.[5] Sulekha.com, founded by Satya Prabhakar and Sangeeta Kshettry as a platform enabling different forms of monetizable interactions among Indians, raised its initial investment from Indigo Monsoon Group and, later, from the Palo Alto–based venture capital firm Norwest Venture Partners. It is today best known as a web-based search engine and "decision-making platform" for semi-organized, gig-economy local services in India, aggregating databases of service providers that include entries for home care, computer training, service apartments, party catering, babysitting, yoga lessons, and auto repair (similar to Checkatrade.com in the UK). The Sulekha app contains options for both standardized local needs (like pest control) and special requirements (catering or interior design).

The ad is set in a nameless Northwest India, perhaps Rajasthan, Maharashtra, or Gujurat, and the aesthetic is neo-indie, low-budget Bollywood comedy (e.g., *Finding Fanny, Pipli Live, Hera Pheri, Ishqiya*); a lilting, festive Indi-pop song gives an upbeat rhythm to the scenes and its editing. Nikhil (Nitin Ratnaparkhi), in his late thirties, bears all the marks of a lower-middle-class shopkeeper or bookkeeper—he is Hindu, seemingly educated in "Hindi-medium" (i.e., studied in a nonglobalized local dialect), hair neatly combed, dressed casually but sensibly. He stops his modest motor bike by the side of a village road where four mixed-generation, "traditional" rural women are trying to hail a ride. They skeptically size up Nikhil's ride, and turn away in frustration: there's four of them and only one seat. The jugaadu's "light-bulb" expression flickers on through a close-up, and suddenly we cut to the four women sitting precariously and anxiously on a sofa strapped to the back of the moped; Nikhil grins as he drives them all away. Cut next to a hot and muggy restaurant, where two men fight over the direction of the ancient water-cooled fan, both wanting to provide comfort (*sukh*) to his family; Nikhil, our man jugaadu (pronounced *joo-ghar-roo*), springs to life, takes off his polyester slacks, and attaches them to the cooler, suddenly giving both families their very own stream of air, one from each pant leg. One of the families leaves, seemingly in disgust. But Nikhil once again grins from ear to ear. Then we cut to the scene of a "kanya dhaan" (the marriage ritual in which the bride leaves her father's home). Nikhil is marrying Urmila, when one of the back tires on the car that is supposed to whisk the two away punctures and breaks. The jugaadu worries through a momentary lapse in confidence, but then the light-bulb expression flickers again, and we cut to the married couple driving away, as the tire has been impossibly replaced by a cart's wheel and axle. Urmila's father throws flowers at them in disgust and worry. The couple is greeted at home by Nikhil's family; Urmila, the new bride, quickly spies all the decrepit, cost-saving fixes that Nikhil has deployed to keep his home in some semblance of hanging-by-a-thread order (uneven plastered walls, stacks of old books for cabinet legs, a repurposed metal pail as a showerhead, etc.). And then the lights go out (at which point the music scratches off, as if the turntable's energy were cut). However, this is only an occasion for the jugaadu's internal light bulb to flicker on: hacker time again. Nikhil rewires the flat's lights through his moped, but on starting the motor, not only does the electricity blow out but also the water main busts (municipal water and the electricity grid are the two key "public" utilities that separate legal from illegal homes). His new bride says without ceremony, and in colloquial Hindi, "Look, don't show off all these jugaads in my home; otherwise I'll install someone else." She takes out her smartphone and opens the Sulekha app, and we see a final cut to a close-up of Nikhil, fearful and worried. The voice-over says, "Sulekha. Just click and get reliable service partners who understand that work doesn't happen through jugaad. Sulekha: Go Anti-Jugaad!"

Throughout the ad, a smooth-voiced man lilts carefree lyrics to an upbeat, guitar strummy, Indi-pop theme song ostensibly in celebration of the jugaadu in us all.

> I go along, giving the gift of peace of mind
> I will fill your life with advantages
> I'll make a broken heart healthy again. . . .
> I can make a flop a hit
> What can't be done, I can do
> I can lift the fallen
> Give rest to the tired
> My name is Jugaadu!

The song humorously screeches to a halt when Urmila rejects Nikhil's jugaad attempts at social reproduction. In Nikhil's impermanent if frugal world, makeshift technology is continuously facilitating and sometimes disrupting the flows connecting his ecology to local and global feedbacks (pirate and monopolistic logistics) and feedforwards (e.g., preemptive and algorithmic control). In each instance, his jugaads are trying to "help gendered subalterns"—the women hitchhikers, the wives and children of the brawling men, and finally Urmila. For her part, Urmila seems to articulate brand Sulekha's sustainable orientation to the domestic sphere: get it fixed professionally, or risk the home space collapsing. There is an ambivalence here as her demand comes from the empowered and networked women jugaadu of the heterosexual home, as well as from the desire to have a formally organized and managed space for social reproduction. The sense is that the jugaadu's seeming concern for the suffering, gendered Other is naive, selfish, and superficial. Nikhil's miserliness and his autonomous technological fetish must be domesticated, his jugaads overturned by the table of values of elite consumption. Commenting on the theme of the nudge campaign, Tithi Ghosh, a senior vice president and head of advertising at Ogilvy and Mather, articulated the cornerstone of the campaign as "the thought to go anti-jugaad. The inconvenience and the pain involved [in finding] a suitable service provider lead[s] to procrastination and temporary, imperfect fixes or makeshift solutions. We decided to use the very Indian cultural phenomenon of jugaad as the springboard for the creative. By dramatising the ill-effects of jugaad at home, we deliver the message that home owners can avail expert help on Sulekha" (Anon. 2016).

Now, jugaad's intimate if uncomfortable relation to neoliberal forms of measure and risk mitigation should be carefully diagrammed. These capitalist values articulate a chain of common sense central to the dogmas of the neoliberal free market: consumption, convenience, satisfaction, profit, creativity, relative surplus value, time management, gig economy, monopoly ownership, spectacularizing risk, and digital crowd-sourced solutions. Already, a fairly wide but distinct range of sense-making is legible in jugaad discourses performed in India today. The digital and the mobile are key to the "perverse implantation" of neoliberalism. Thus, Soumendu Ganguly, the head of marketing at Sulekha.com, says, "Sulekha is one of India's largest digital brands. People know about us and that's why we get close to 20 million visitors every month. But, we were considered to be a classified website. With an explosive increase in India's digital population, upwardly mobile Indians are looking online for all their needs. Sulekha being an early mover in this category with its vast network of trusted service professionals, wanted to appropriate this space by going on mass media and claiming the category" (Anon. 2016). Sulekha offers differential access to the elite, upwardly mobile consumption of services in India; it is operating on the model of Uber, AirBnB, Gumtree, and other crowd-sourced digitally networked service-gig aggregators; indeed, these kinds of networked, sharing economy platforms are themselves understood as kinds of digitally networked jugaads. At Sulekha what's on offer are trust and professionalism legitimated through an aggressively managed digital media organization.

Ganguly further adds, "We believe the insight for the campaign is deep-rooted in Indian culture and would strike a chord with the audience. We have all tried to jugaad our way out of situations. While it is a popular practice, we all know that it is not the optimum solution. Local service partners listed on Sulekha understand that consumers aren't looking for quick fixes, but for permanent solutions. Our 'Go #AntiJugaad' campaign echoes this sentiment and aims to reach out to everyone who seeks easy access to professional, high-quality local services" (Anon. 2016). Why anti-jugaad? If on the one hand contemporary Indian media momentarily nominates jugaad to be the essential characteristic of Indianness, on the other it is seen as a shameful national habit, a kind of miserly shopkeeper's make-do ethos.[6] The ironic recourse to deep-culture (i.e., essentialist) marketing refers to and obscures other relations: How transparent is the system of measures determining quality? What situations call for jugaad and which don't, who gets to say so, and on what basis? And what exactly, given the evolving ecologies of matter, population, policy, capital, land, water, identity, and biological life, is a permanent, optimal solution? Intensively and extensively today in neoliberal India this solution involves the mobile phone. Satya Prabhakar, the CEO of Sulekha.com, feels that smartphones have ushered in a new era of local services in India, and Sulekha ideally wants to tap in to it with the campaign. "Smartphones have revolutionized how Indians search for and consume local services. It has become an important business category and the market is currently valued at $200 billion. At Sulekha, we have had a 90 per cent growth in local services demand last year, and it will only grow. We are committed to make this business a success" (Anon. 2016).[7] The pedagogy of elite consumption passes through the pirated prism of the mobile phone, and a branded network of service providers will spread a new ethos: anti-jugaad will win the day. But what are we to make of the moment of jugaad "innovation" in relation to affection and affect?

Indeed, what is at stake in these sets of representations, these affective images? In affective ethnographies, affirmation and ethics pass beyond good and evil to consider material infrastructures of good and bad, that is, the fuzzy set of joyous passions as emergent quasi-causes of gradients in an ecology's morphogenesis. These fuzzy sets of sensation, in their mode of ontological affectivity and epistemological common notion, bring to crisis both Western humanism and the regime of what Elizabeth Povinelli (2016), following Quentin Meillasoux, has called "correlationism."

For Povinelli, a common thread connecting the diverse schools of speculative materialism is a shared abhorrence of Kant's influence on metaphysics. But, as she is careful to point out, many differences separate the schools. "Thus, if Meillassoux's approach is to demonstrate that humans can think the absolute, then Steven Shaviro's solution for how to sidestep the correlationalist trap is to intervene in how we think about thought, affirming his call for a new 'image of thought'" (2016, 125). For Povinelli and Steven Shaviro, thought is not, after all, an especially human privilege, and it is in fact one of the driving insights behind panpsychism. Drawing on recent biological research that seems to indicate that aspects of what we commonly understand as thinking—or an experiential sensitivity to affect and be affected—"goes on in such entities as trees, slime mold, and bacteria, even though none of these organisms have brains. Other forms of existence might not think like humans think, namely apprehend through the semiotic forms of human cognition (categories and reason). But that does not mean they do not think. It means we should think about thinking in another way. A noncorrelational approach to thought—pulled from Charles Peirce's model of the interpretant or George Molnar's concept of aboutness—seems to exist in all things. Advancing a model of thought that would include nonhuman thought 'means developing a notion of thought that is pre-cognitive (involving "feeling" rather than articulated judgments) and non-intentional (not directed towards an object with which it would be correlated)'" (Povinelli 2016, 125–126; see also Shaviro 2014, 14–20). Rather than miring oneself in a philosophical contradiction, thinking how objects can be let to be without human thought transforms first philosophy into aesthetics, which, critically following Whitehead and Graham Harman, Shaviro argues involves a method of turning oppositions into contrasts.

For her part, while Povinelli is sensitive to the question of attention, she relates its processes in their ontological orientations to nonhuman becomings. She does so through the discursive construction of sets of memory-images circulating in the lifeworlds of Northwest Australian aboriginal peoples. Through it, she draws out a resource for what I am calling affective ethnographies, and that is a practice of decolonizing attention.

I turn here to the work of Henri Bergson, Gilles Deleuze, and Félix Guattari for aid in forming a pragmatic notion of jugaad, or more precisely a common notion (i.e., a notion common to two or more multiplicities), which the practice of jugaad affirms in its various assemblages, sensory-motor circuits, and (de)habituations and virtuosities (Deleuze 1988b, 53). As will be clear in the chapters that follow, I have been affected by jugaad practice to develop a wariness of its political ecology. I want to summarize and conclude the discussion of the Sulekha ad by considering its affective relations in the difference between affection and affect, and between intellectual and manual labor. Keep in mind the repeated "innovation"

FIGURE I.2. Nikhil's "jugaad time" face in the Sulekha ad.

image: the light-bulb flickering on jugaad time for Nikhil and his relations is strictly speaking the image—already past but still incomplete—of a passage from one state to another.

Deleuze's work on affection images, taken from his encounter with Spinoza, Kant, Lacan, Nietzsche, and Bergson, allows us to better pose the question of hacking today. First, we have the body's affection and idea involving the nature of the external body, and, second, we have the embodied power of action or affect. Deleuze defines the latter as an increase or decrease of the power of acting, for the "body and mind alike" (1988b, 49). Notice then that the body's affection differentially affects both the mental and manual labor divide; I will return to this in the conclusion to this chapter. Thus, on the one hand, affection refers not to an idealized conception of the body but to a definite state, composition, or set of dispositions of the affected body (the body's "polyphased space," or, as Felix Guattari defines it, an "abstract space where the axes represent the variables characterizing the system" [Guattari 1995, 97; see also Combes 2013, 4]). Fugitive and incomplete, the variables of a body's multiple phase space, seemingly captured in the face's expression of innovation, are precisely what is potentialized in the jugaad practices dramatized in the Sulekha ad (Bergson 2012, 260). But the body's phase space implies the effective action of an affecting body. As feminist work on biopolitical affect has shown, the valence and politics of affect cover a vast range of referents, processes, and struggles (Ahmed 2007; Ashcraft 2017; Berlant 2016; Clough 2010; Cowen and Siciliano 2011; De Angelis 2007, 2010, 2017; Grosz 2013; Knights 2015;

McIntyre and Nast 2011; McLean 2014; Morrissey 2011; Povinelli 2016; Pratt et al. 2017; Puar 2017; Pullen and Rhodes 2015; Pullen et al. 2017; Yusoff 2017; Zeiderman 2018). Emotion, our common sense, but not some universal experience of fear, for instance, is one dimension of affect: it refers thought back to a preindividual and historical form of the bodily capacity to sense and act, and to make sense of that act. But it remains the latter form, which is already the idea of its form (what image of thought is necessary to think affective capacities as well as what Povinelli calls geontology?). The senses of nonhuman and human ecologies within and against semiocapitalism that Povinelli's nuanced analysis activates as a specific postcolonial archive, affect and participate in what is now a general displacement of the postcolonial critique of representation (Povinelli 2016, 215; Derrida 1998; Berardi 2009a). Here, affect refers to the durational passage from one habituated state to another, taking into account a strictly unpredictable "correlative variation of the affecting bodies" as shifting and emergent capacities along gradients of intensity.

There is therefore a difference in nature between affection and affect. An affection-image is a state of composition of the body, and affect follows from it "as from its cause." But, and this is decisive for a thoroughgoing displacement of what Deleuze disparagingly calls Platonic representationalism, the affect "is not confined to the image or idea; it is of another nature, being purely transitive, and not indicative or representative, since it is experienced in a lived duration that involves the difference between two states" (Deleuze 1988b, 49). So affect is the durational difference (following Gabriel Tarde, the identity of a duration would here be simply the most minimal degree of difference [Tarde 2012]) or intensive variation in a body's phase space, while affection is compositional. Intensive difference passes through the durations of affect such that it both repeats and potentializes the compositional state that is the body's affection (Shaviro 2014). Deleuze shows that an existing mode (body) is thereby defined by a definite and yet plastic capacity to affect and be affected.

A mode's essence is a power; to it corresponds a certain capacity of the mode to be affected. But because the mode is a part of Nature, this capacity is always exercised, either in affections produced by external things (those affections called passive), or in affections explained by its own essence (called active). Thus the distinction between power and act, on the level of modes, disappears in favor of two equally actual powers, that of acting, and that of suffering action, which vary inversely one to the other but whose sum is both constant and constantly effective. Thus Spinoza can sometimes present the power of modes as an invariant identical to their essence, since the capacity to be affected remains fixed, and sometimes as subject to variation, since the power of acting (or force of existing) "increases" and "diminishes" according to the proportion of active affections contributing to the exercise of this power at any moment. It remains that a mode, in any case, has no power that is not actual: it is at each moment all that it can be, its power is its essence. (Deleuze 1992a, 93)

Povinelli, in her creative engagement of the notion of essence as power, notes that the "power (potenza) of potentiality is the positivity within bio-power, within Life" (2016, 80; on essence as power see also Hardt, 1995; Deleuze 1992a). Following Shaviro (2014) and Jane Bennett (2010), Povinelli argues for a break from a focus on essences in affective ethnographies. "When the focus of the ontology of self-organized being is shifted from the search for essences to the desire for events, from sharp epidermal boundaries to fuzzy and open borders, and from simple local bodies to complex global patterns, the following emerge as exemplary ontological objects: weather systems, carbon cycles, computer routing systems" (2016, 46). This movement away from epidermally enclosed, self-oriented, and self-organized entities and toward the complex dynamics of far-from-equilibrium assemblages likewise characterizes Bennett's model of a post-biopolitics grounded in the concepts of actants, affects, and events rather than in the processes of life differentiating from nonlife. We will return to the potentiality of technoperceptual plasticity in the chapters that follow, but here I highlight its relation to the affective event.

And here too we are on the experimental terrain of composing an affective method. When one mode encounters another mode, an affect-event may be produced in which an operation of resonance takes hold of both modes, such that both modes enter into a third relation, a relation with the outside of what folds them together in composition or decomposition (Culp 2016; Combes 2013, 36; Massumi 2002). The light-bulb effect, the sensorimotor circuit of what, in business and management discourses as we shall see, is celebrated as creativity and innovation itself, is in Sulekha's (and behind them Ogilvy and Mather's) elite appropriation and rejection of the "deep" culture of jugaad, the result of the high-risk habits of the "incorrigibly" precarious. Affective method, then, involves a thinking and practice of essences as ontological and epistemological at once (Barad 2007; Deleuze 1992a). Power as (in)capacity, to think with both Andrew Culp (2016) and Jasbir Puar (2017), becomes the affective-image and common notion of an essentialized jugaad, its specific regime of potentialization, habituation, and capture.

The affect of jugaad is bound up with what Sohn-Rethel analyzed as one of the fundamental problems of capitalism: the division between intellectual and manual labor. Now, if we take jugaad as method, that is, as the idea of the idea of working around to the point of sabotaging what's given as fixed, normal, formal, propertied, *suvarna* (upper caste), appropriate and right, its joyous practice destabilizes not only the value-form of commodity production for monopolistic control, but also the enforced dichotomy between intellectual and manual labor in several caste and class hierarchies that it presupposes. Sohn-Rethel links this latter dichotomy to the social synthesis produced through the abstraction of money (1978, 6).

Clearly the division between the labour of head and hand stretches in one form or another throughout the whole history of class society and economic exploitation. It is one of the phenomena of alienation on which exploitation feeds. Nevertheless, it is by no means self-apparent how a ruling class invariably has at its command the specific form of mental labour which it requires. And although by its roots it is obviously bound up with the conditions underlying the class rule, the mental labour of a particular epoch does require a certain independence to be of use to the ruling class. Nor are the bearers of the mental labour, be they priests, philosophers or scientists, the main beneficiaries of the rule to which they contribute, they remain its servants. The objective value of their function, and even the standard of truth itself, emerge in history in the course of the division of head and hand which in its turn is part of the class rule. Thus objective truth and its class function are connected at their very roots and it is only if they can be seen thus linked, logically and historically that they can be explained. But what implications does this have for the possibility of a modern, classless and yet highly technological society? This question leads on to the need for a further extension of Marxist theory which did not arise at an earlier epoch: what is in fact the effective line of differentiation between a class society and a classless one? . . . The three groups of questions raised here

> stand in an inner relationship to each other. The link connecting them is the social synthesis: the network of relations by which society forms a coherent whole. . . . As social forms develop and change, so also does the synthesis which holds together the multiplicity of links operating between men according to the division of labour. (Sohn-Rethel 1978, 4)

The deconstruction of binaries (e.g., the reversal and displacement of manual and intellectual labor) has become a mechanical process—Facebook and Google bots do it regularly. That is not my aim here: material *gradients* between intellectual and manual labor pass through the affective dynamism of historically situated bodies whose power, whose essence (power) we do not fully know. In the chapters that follow, I will return to the dichotomy highlighted in Sohn-Rethel's study of class formation; here I want to mark how the affective passage from obstacle to flow in the Sulekha ad is tied to (at least) two movements at once: the affective, durational, and embodied passage from one state of bodily affordances to another, and a general, but again embodied displacement of the dichotomy of manual and intellectual labor. Power's (in)capacity and creative labor: do these two movements resonate and/or intersect, and are these abstractions—resonation and intersection—merely two expressions of a certain will to power within postcolonial practices of affective ethnography (Culp 2016)?[8]

The Sulekha ad humorously stages the class and gender drama of jugaad as the discourse and practice of India's new "extreme work" cultures (Bloomfield and Dale 2015; Gascoigne et al. 2015), in which the work-around becomes work. In the ad, the song highlights the euphoric egoism of the jugaadu: the light bulb *chamatkar,* the miracle or marvel of innovation as intellectual labor of an entrepreneurial service class, is immanent to Nikhil, and its benefits (through manual, but highly stylized labor), its affective passages, are self-evident (to him). The frugality of the jugaadu poses a radical, destabilizing, and unsustainable risk to social reproduction, as water and electricity disruption stops up all other flows—from affection to shitting to money. Jugaad is an enemy of sustainability, and so it is an enemy of the (gendered and sexed) home (and through association also anathema to true religion); it is an obstacle to security and trust. David Harvey has usefully noted that the

> monopoly power of private property is, therefore, both the beginning and the endpoint of all capitalist activity. A non-tradeable juridical right exists at the very foundation of all capitalist trade, making the option

of non-trading (hoarding, withholding, miserly behavior) an important problem in capitalist markets. Pure market competition, free commodity exchange and perfect market rationality are, therefore, rather rare and chronically unstable devices for coordinating production and consumption decisions. The problem is to keep economic relations competitive enough while sustaining the individual and class monopoly privileges of private property that are the foundation of capitalism as a political-economic system. (2002, 97)

As we have seen in the Sulekha anti-jugaad "campaign," class, gender, ability, and religion are articulated in a whole ecology of sensation, continually synthesizing feedbacked perception with machinic capacities. One might say that a kind of class-caste war is being conducted through the pedagogy of Sulekha's anti-jugaad: "Elevate yourselves from the manual pettiness of your backward caste precarity," it seems to command. It is based on the radical separation between jugaad as vulgar, low-caste, manual, amateur, material, creative, DIY, and disorganized/informal practice versus professional, pure, suvarna, formally organized, value-added, insured, and networked labor (Athique et al. 2018, 10–15). Yet, in practice, there is no separation: as we shall see, makeshift, heterodox economic practice and logistics, and mobile media workarounds are differentially meshed throughout the precarious formal and informal ecologies of social reproduction in India's so-called smart cities. Thus, we come to the image of our first approximation: not Nikhil, but his ad hoc, preindividual ecologies, the dynamics of egoistic affections, joyous passions separating the ecology from what it can do, expelling solidarity for sympathy, the effects of unconscious heterosexual masculinity (the hacker-engineer as the active, joyous man) and a vague, backgrounded Hinduism, the embodied, material flows that give the lie to any rigorous separation between formal and informal economies as well as intellectual and manual labor—these processual dynamics, differently crosshatched in each sign, close-up, and shot-reverse shot, are usually obfuscated through the commodification, gridding, and capture of the affective relations of the jugaad image (see Deleuze 1994). While, as I will show, jugaad practice is embraced enthusiastically by both men and women, its majoritarian media representation usually puts its practice in the domain of masculine control, rarely in terms of women's joyous and virtuosic hacking into the conditions of their own exploitation.

The Jugaadu's Smart City

In chapters 3 and 4, I turn to the problem of hacking the smart/Big Data city. Here I want to pave the way for that discussion by asking another question in the context of a short documentary on subaltern media ecologies: How does postcolonial, urban media practice in India refocus the question of the political today? As I suggested above, the embodiment of media in India returns us to the project of a political ecology of the image. Recall that for Bergson an image is halfway between a representation and a thing, and if it has a "life" it is through the temporal and material organization of noncoinciding resonant unities, or moving wholes, which could be durations, a neural network, the murmuration of a flock of birds, or assemblages of assemblages of as many other things besides. In diagramming practices of the image as sensory-motor circuits in ecological feedback with assemblages of matter, bacteria, minerals, speed, and technologies (all with varying degrees of force and vectors of change for a given assemblage), what is at stake for the question of a method of experimentation in affect? I will offer some introductory thoughts on the question of method in my final section of this chapter, but here I want to turn to the collective assemblages of these experiments and consider their ontologies. One of the great challenges of Deleuze's work on affect is to resituate the question of the political ecology of capital ontologically, through what others have called variously ontopower, necropower, biocapital, or geontopower (Deleuze 1986, 1988a, 1988b; Massumi 2015a; see also Manning 2009, 2013; Manning and Massumi 2014; Mbembe 2001; Povinelli 2016). Hence, if affect is autonomous, one line of flight for this multiplicity is the political itself. This is not, however, to blithely affirm in the manner of a braying ass that can only say yes some vague horizontal connectivism that would bestow a technoperceptual joy onto subaltern struggles against regimes of neoliberal debility. As Culp (2016, 17) warns,

> Joy surfaces as the feeling of pleasure that comes when a body encounters something that expands its capacities, which are affects said to "agree with my nature," to be "good" or simply "useful." To end the story here (though some do) would reproduce a naive hedonism based on inquiries into subjects and their self-reported affective states. Spinoza's theory of affects is not an affirmation of a subject's feelings but a proof

of the inadequacy of critique. Affects are by-products emitted during the encounter that hint at a replacement for recognition or understanding as the feedback loop to indicate if knowledge was sufficient. But there are innumerable forms of knowledge, many of which invite stupidity or illusion.

The importance of this warning affects a shift in affective dispositions, from parataxis to parataxis, through the diagrammatic method. The "darkness" that Culp aims to raise to a higher power and purpose (2016, 32), an untimely hatred of biopolitical capital and all that flows from it, suggests that the diagrammatic method poses questions of the passions (joyous and sad) precisely to develop common notions traversing multiplicities and the stupidity that captures them. In what follows, I diagram the autonomy in affect through a consideration of jugaad subjectivity in a documentary film about media piracy in Mumbai, *Videokaaran* (2012).

Jagannathan Krishnan's *Videokaaran* begins with a scene of a get-together of a few working-class, Dalit male fans of Indian cinema. Later in the film this clip is resituated through a recursive unfolding of a vibrant, largely male subaltern media ecology. The "hero" of the film is a member of what he refers to as a "criminal" network (with its own don!); he is also active in the Christian community. Sagai Raj, a thirtysomething tenth-standard pass (basic primary education), is a media entrepreneur in a slum of Bombay. He once owned a video parlor (popular urban exhibition and social spaces screening video CDs to working class and Dalit communities; see Titus, forthcoming) but now runs his own photography studio, and on the side he helps run a porn-video-smuggling network through assorted video piracy practices.

The clip shows Sagai sitting in a darkened room with his friend (and the filmmaker) discussing the "janoon" (madness) of Indian cinema. He discloses that his own connection to cinema (and to kriya yoga) is through Rajnikanth, the popular Tamil film star. He immediately contrasts him to Hindi film star (and brand) Amitabh Bachchan, hoping to draw his friend out to deliver some famous dialogues. The "somewhat forced" conversation turns to Rajnikanth and Bachchan's costarring film *Hum* (with Govinda), which revitalized Amitabh's flagging career back in the early 1990s; in it they had some chemistry, they all agree. Sagai refocuses attention on a comparison between Rajnikanth and Amitabh; he contrasts their trademark entries into films in terms of their speeds (fast and slow, respectively)

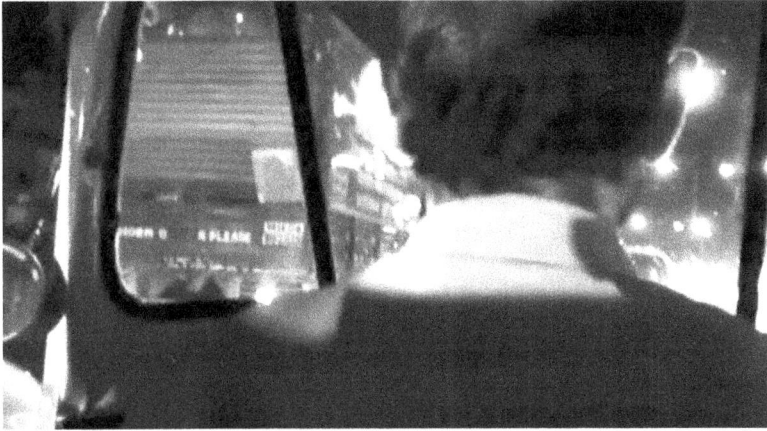

FIGURE I.3. Sagai Raj, nighttime vision shot.

and what qualities of the actor's style they allow to be emphasized (action versus dialogue).

It's a shrewd observation: Rajnikanth's sonically weaponized and lightning-fast gestures are well-known signatures of his style. Throughout *Videokaaran*, the viewer pieces together a precarious ecology of image, desire, movement, media technology, class/caste, masculinity, intellectual property, and law (Berardi 2008; Berlant 2016; Butler 2006; Hardt and Negri 1999, 2001, 2009; Harvey 2002; Liang, 2009; Lovink and Rossiter 2007; Sundaram 2009; Terranova 2004). Through it we glimpse the traces of social struggles to common resources that are trapped in the monopoly rents of private property. In an urban dialect of easy misogynist masculinity (mothers and sisters figure heavily, with scatological swear words punctuating each sentence), the entrepreneurs of affective style discuss the gentrification of cinema—ticket prices certainly (talkie versus multiplex [Ganti 2012; Rai 2009]), but also the destruction of subaltern video-parlor culture due to copyright policing and corrupt licensing. This was Krishnan's initial intuition in making the film: the cinema is being taken away from the poor (others have documented how the poor have been taken out of dominant Hindi cinema—see Ganti 2012). The demolition of the video parlor, in a strange but unintended haunting of the demolition of the Babri masjid in Ayodhya that sparked the nationwide resurgence of Hindu chauvinism in 1991, is one act in this history of the gentrification of cinema in India. The mise-en-scène is almost claustrophobic, with extreme close-ups of mouths laughing with shiny, yellowed teeth; gray shapes against a vaguely glowing suburban night; greenish nightvisioned ghosts; slips of an unsteady handheld camera glancing over naked male torsos; missing actants (a woman who is repeatedly named but only fleetingly filmed—the drama of colonial ethnographic filmmaking haunts the mise-en-scène); the dangers of visibility for the ones who follow the arcane path of jugaad, blurring TV screen shots of movies; reaction shots of Sagai watching TV (sets of sensory-motor circuits); and the social and economic daring of living in poverty.

What image is this? Is that a well-posed question today in India? Deleuze's cinema books develop a typology of Bergsonian images as they circulate through and create sensory-motor effects. An effective history of affect is centrally at issue in a striking early passage in *Cinema 1: The Movement-Image* (Deleuze 1986). This image concerns time-as-duration, but also an artistic practice that experiments in the affects of the interval of durations, an image that would be relegated to the humanism of the dialectic in Deleuze's subsequent analysis of Soviet montage, but one that points to a way of encountering what Sagai Raj expresses in *Videokaaran*. This is the pathos-image, which is not sadness but rather some kind of mixture of intuitive, immanent firstness and relational secondness (Pierce 1995; Deleuze 1986, 98), images that flash out untimely zones of indeterminate intensities and nonlinear processes. "Kya bolunga main? [What can I say?]," Sagai Raj asks filmmaker Jagannathan through a face that conjoins the intensity of cinephilia with the recognition of speaking to someone (the filmmaker) who does not know Rajnikanth in Tamil.

Deleuze speaks of the pathetic image as involving two aspects. Recalling our discussion of affection and affect in the Sulekha ad, here we see simultaneously the transition from one term or quality to another, and the sudden emergence of a new quality that is born from the transition accomplished: the pathetic image is both compression and explosion. Deleuze writes of the acceleration of qualities moving through the movement of the montage, and in so doing the image passes to a "higher power" (1986, 35), or a passage into a new dimension, raising it to the $n+1$ power. What are the implications of this accelerated morphogenesis into a new dimension? It is the interval within jugaad ecologies, as affective passage (Deleuze 1988b), which now takes on a new meaning: the interval is the qualitative and abductive leap into the raised power of the instant.

I take Deleuze's treatment of the pathetic as part of a certain proposition on the ontology of affective intervals. Not unlike Nikhil in the Sulekha ad, with Sagai Raj we see reiterated the expression of a certain style of entrepreneurial masculinist piracy/hacking; in this image, the manual and intellectual labor of invention is displaced in and through a subject who, living partly in jugaad time, partly in actuality, seems slightly out of kilter, a bit odd, a "geek," and at his best a kind of single-minded virtuoso of the hack, not untimely, perhaps, but hypertimely. He narrates in breathless yet measured Bombay Hindi his many attempts to produce and capture value in and through media—as image maker for others, wedding video maker, Photoshop expert, curator of a machinic assemblage of interpolated bits of porn, as a designer of a largely masculinist social center (i.e., the video parlor) for the exhibition of the pan-genres of Tamil cinema. He demonstrates the parlor's logistics of recirculation, its existence against surveillance, the litter of the Anthropocene, and the naturalization of privatized property (Harney and Moten 2013). The video parlor is a space of commoning and extracting (Barbagallo 2015), or sharing and exploiting; the organizational practices include watching for police, using decoys and costumes, directing traffic in and out of the video parlor away from the train tracks, weed and liquor consumption, cameras, communications, and image-production. These curation practices undergird and/or facilitate piracy ecologies. The director's camera has come after the catastrophe of the demolition of the video parlor—the camera both commemorates and circulates this disaster: turning the device that records the memory of this injustice, Sagai Raj shoots the director trying to light a cigarette (Jagannathan had originally intended to hire Sagai to shoot the movie, but quickly realized it would be better to have Sagai as its hero).

I think of this anomalous documentary in a sense as speaking to the world described so well by David Harvey, that world in which capitalist rent is an art. *Videokaaran* tells the story of those who practice another quasi-capitalist art: jugaads, workarounds, life hacking, or virtuosic precarity. In this world, jugaads produce value and profit, but usually at a very small scale, and that too momentarily; meanwhile the jugaadu, who, through his canny displacement of the dichotomy between head and hand, has unwittingly become the vehicle of a kind of preindividual fetish for jugaad, and thereby accumulates reputational value and, typically, money. In the giddiness of Sagai Raj's description of his extralegal escapades is an ecstatic embrace of the necessary contingency of movement and action in the formal capitalist and informalizing hacking ecologies that work his assemblage. But already Sagai Raj takes the question of media practice to a higher level, beyond just the vagaries of the workaround, and he raises it to the level of a philosophy of virtuosity, which is also a philosophy of technique, or pragmatics. Sagai Raj celebrates his criminal activities as a style of creating images, for others, for himself. Both his activities and his images take him into the realm of the pathetic: not sadness, but into the preindividual potentiality of affect, in rhythmic motions of compression and explosion. Sagai Raj reads the signs of affect. He tells the director, I can read your face and body and know your presence before you approach me. "Sagai analyses the behaviour of policemen, and studies people so closely that 'even when I look at a shadow I know who it is. When we were screening films we had to monitor the audience and be alert all the time.' He and his friends have been so influenced by movie stars that they are already natural performers—the swagger and the smart lines come easily to them."[9] In one scene, Sagai describes the best strategy for taking a beating from a cop. After simulating a particular threshold of pain, he crumples his body, expressing and dissimulating a physical limit. Indeed his image streams multiply expressions; his different practices have their own but sometimes overlapping image domain. In another, folded intimately with his admiration for film stars, he describes stalking women for sex: not sex workers, but women who he knows "want it." When queried about the seeming misogyny in his objectifications, he literally shrugs it off.[10] These contradictory foldings of affects, habits, memories, and violence show that in subaltern media ecologies whose own domain is organized through increasingly arcane and highly stylized arts of the extralegal and of the exception we are witness to important laws of capitalist accumulation and masculinist violence (Agamben 2005; Roy 2009). Property is expropriation; expropriation is violence; violence is the law of suvarna capital. Meanwhile, the pirated is made common; the commons informalize capital; the commons are the future anterior of capital. There is no symmetry or dialectic here, however. The passage from law to jugaad to the ethics of the extralegal and the commons involves the analyses of nonlinear rhythms, the effects of which are sometimes indiscernible yet real, insofar as they are virtual. These virtual, abstract rhythms, as joyous passions autonomous, preindividual, quasi-casual and in relation, link the speeds and gestures of favorite masculinist heroes to the cycles, violence, and pleasures of the extralegal. This involves at times state-sanctioned entrepreneurship, marking Sagai Raj's career—negotiating beatings by the police while working the pirate kingdom to his own (gendered) advantage.

Harvey's work opens another question: the monopoly advantage of both intellectual property and first-mover advantage is the source of license rents that overwhelmingly favor oligopolistically configured multinational corporations based in the global North, and intense sites of social and political struggle throughout the world (Harvey 2002; Nolan and Zhang 2007; Smiers 2007). For Jagannathan recounting his experiences making the film, copyright did not figure as a major obstacle in Sagai's media practices (personal correspondence, April 7, 2016). Indeed, intellectual property in today's distributed piracy kingdoms (Sundaram 2009) is increasingly seen as merely a historical phase of organizing and controlling the creation, distribution, and consumption of experience (e.g., in the creative industries or smart cities scheme). However, in the economies of the global South low wages, desperately precarious agricultural conditions, urban and factory-related ecological disasters, chronic water shortages, and poor access to poor infrastructures continue to characterize everyday life (see Amin and Thrift 2002; Birkinshaw 2016; Gill and Pratt 2008; Hesmondhalgh and Baker 2010; Smiers 2007). India has been unevenly integrated into the service and creative industries of the global North, and recent policies by both the former Congress and present BJP-led government have sought to encourage "glocal" brand presences for Indian companies. Going by the hype of the current BJP chauvinism and their plans for smart cities, the future seems to be blisteringly bright for India's creative industries.

But Sagai's pirate kingdom is the excluded center of India's Acche Din (good days), as Prime Minister Narendra Modi calls our era of falsely imprisoned students, censored histories, neoliberal corrupt rule, heterosexist stigmatization, lynch law, murdered journalists, murdered Dalit intellectuals, and might-is-right politics. The hacking ecology that Sagai mobilizes to jugaad his way to his next image show is a state of exception, a necropolitical domain, with its joyous and sad passions that continue to separate its own ecology from what it can do. Its intensities and events in forms of life and politics in poor and Dalit urban communities suggest another dimension to the passage of the interval, but one that composes political subjectivities and affections that no longer bear clear connections to older party-style affiliations or even caste, class, and religious identities. *Videokaaran* does not shy away from the pervasive presence of the conservative Hindu Shiv Sena throughout Bombay life—the final scenes are of Sagai Raj, himself a Dalit Christian, walking through a Hindutva crowd in the midst of a boisterous procession, seemingly both within and outside this other masculinist assemblage. The extralegal image of the jugaadu (as virtuoso of the workaround) does not disclose a politics, but the conditions of possibility of politics as such. Those conditions suggest that media and affect are twined in the body, compressing and exploding habituations continually. *Videokaaran* brings out other ecological relations in jugaad practices. Affect in jugaad time, as the durational passage from a static affection to experimentation, triggers euphoria for Sagai Raj; his at times docile, at times aggressive masculinity coevolved with his machinic assemblages and media consumption practices.

To summarize and conclude, jugaad, as I will argue in the chapters to come, is an image of entrepreneurial abandon and virtuosity, one that curiously withdraws value from its otherwise accumulating and circular flow (Schumpeter 2008). The intensive parataxis draws forward by juxtaposing contrasting if not contradictory jugaad images:

1   The social practice of jugaad is aligned with what is essentially Indian, and it is excoriated as incorrigibly premodern. Jugaad, thus, brings out the questions, Which essence? Which modernities?

2   The social practice of jugaad is celebrated as innovation, and it is decried as value-blocking. In this and many other ways, the value of jugaad involves the political economy of nonformalized creativity in India today.

3   Jugaad is accepted as the law of precarity, and it is moralized against as short-sighted egoism. As subjectivity, jugaad involves the materiality of class, caste, and privilege and consumerist habituation.

4   Jugaad is celebrated as the new rule for frugalizing India, and it is bemoaned as uncontrollable extralegal "exception." Law and poverty collude to keep jugaad a joyous passion, strategically abstracted from but feedbacked to 1, 2, and 3.

The ecology of these contrasts constitutes different ontological dimensions of jugaad practices. This is why for both neoliberalism (Sulekha's campaign to domesticate the jugaadu) and the law (Sagai's pirate exception to its rule) the patronage and associations through which jugaads take form ensure an acceptable, probable degree of fluctuations in existing relations of sense, force, and value.

Jugaad as Affective Practice and Critical Method

New critical methods of affecting and sensing the technological substrates of bodily and collective capacities have emerged today. These methods disrupt, repurpose, and/or reassemble molar and molecular relations of force, sense, and value, and they develop critical feedbacks to politico-theoretical practices. Throughout *Jugaad Time*, I consider the everyday practice of jugaad as a potentializing affective passage and intercalated hinge between actuality and virtuality. How does this hinge perform its relations in the processes of doing research? What does this suggest about the geopolitics of research today? Can a kind of pragmaticism of jugaad enable a critical reflection on doing research in affective practices under conditions of neoliberal, postcolonial, racialized, late, all-too-late capital?

Let us begin with a distinction of attention: when we refer to a jugaad event, what is the referent of this term? Jugaad as a workaround, for example, repurposing a brassiere as a motor's belt, can be part of a kind of life-hacking philosophy.[11] However, as a practice it is not solely, or primarily, focused on *hacking* anything; its affective disposition intuits a purposeful, continual, mental, and manual tinkering focused on getting something done, elegantly and beautifully if possible, but, even if in a patchwork, makeshift, even ugly way, done. Mobilizing a mode of attention that Povinelli brings to crisis in *Geontologies*, tinkering-based workarounds can sometimes be direct hacks into forms of power, whether algorithmic or not, but they need not be. With varying intensities of attention, workarounds can open interfaces, massage machines and media, for better, quicker flow of more work, more pleasure, more (self-)exploitation. Workarounds can literally create time out of timepass (on timepass see Rai 2009). The overworked neighborhood mobile-repair adept attends to your faulty device with intuitive, implicit, and formal jugaad diagrams to make the device functional again; its sad and joyous passions reverberate through your habits whose processes express (in)capacities and powers that involve the jugaad in other political economies of becoming.

Jugaad then can be understood through several utility measures, grids of efficiency, and tables of values common to neoliberalizing postcolonial economies (Brown 2015; Povinelli 2016), but specifically developed in the informal economies of India (see Lloyd-Evans 2008). If we consider the implications for both politics and theory of recent work in postcolonial queer disability studies—for instance, the work of Puar (2017) and Dasgupta and Dasgupta (2018)—we might ask what jugaad can offer a method of research. Is there an archive of queer jugaad ecologies from which interdisciplinary methodologies can learn?

In social practices ranging from the virtuosity of mobile phone repair wallahs and the constant repurposing of media devices in informal piracy ecologies, to negotiating the disabling, debilitating Brahmanical biopolitics (onto/geontopower) of caste through different practices of commoning, workaround, and refusal, the affective disposition of jugaad can be linked to what Brian Massumi (2015a, 2015b) has called the priming of ontopower. Ontopower is a form of economic, political, and social power that generates, as we have seen above, affections (as states of being) and affects (as a durational passage from one state to another) preemptively, that is, in anticipation of events, keeping the body's capacity in a state of ever-ready deployment. Jugaad practices anticipate the functioning of ontopower: in its tinkering with material and intensive affordances and continually recalculated strategies interfacing with different combinations of assemblages, jugaad can potentialize conditions of extreme precarity. Simultaneously, jugaad mobilizes tendencies common to the ecologies of sensation of prosumer neoliberalism: individualism, consumption, short-term fixes, award-yielding work, savings, debt, human capital, entrepreneurialism, and disruptive innovation. Here, jugaad, far from being a practice of autonomous hacking, capitulates to and intensifies capitalist habituation and surplus value accumulation.

Indian neoliberalism has developed its own jugaad image, as the practice is enthusiastically taken up in business management discourses, euphemistically refashioned as "frugal innovation." Acting within and against these systems of force, sense, and value, jugaad enables a renewed focus on the unequal material conditions of its ecologies, the forms of struggle commensurate with their overthrow, and the diagrams for a noncapitalist landing to our collective lines of flight (De Angelis, 2007, 2010, 2017). However, the revolutionary becoming moving through jugaad practice suggests we may never be landing as such. Formal sociological categories or images of thought that foreground the primacy of academic value will find in jugaad diagrams an impossible and sloppy eclecticism. Normalizing jugaad through this image of thought, which we can call, following through on some provocative work in organizational studies, the image of excellence within a neoliberal university, yields nothing (Ashcraft 2017; Thanem and Wallenberg 2015; Dale and Latham 2015). The essence of (non)human freedom itself is in play in every jugaad.

A pragmatic and effective political economy of the heterodox practices of informality legible in jugaad practice calls for both a speculative method conjoining virtual and actual affects (affordances, capacities, tendencies, debilities), and an experimentation in and through ecologies that traverse digital-analogue assemblages of contemporary postcolonial capital (Massumi 2015b). In interviews with practitioners, jugaad became for me an attentive and canny bodily orientation toward historically specific *dispositifs* of power, exploitation, discourse, materiality, value, and intensity, *and* a relational practice of experiencing, negotiating, and, at times, changing human and nonhuman ecologies. Many of these interviews were conducted while specific jugaads were being affected; sometimes the interviews themselves were a kind of jugaad for our interlocutors. Jugaad works with, and at times creates, metabolic imbroglios—events in the informal economies of the precariat's biopolitical production of both resistance and dispossession in contemporary postcolonial capitalism in India (Arboleda 2015). Here, I begin considering examples to which I will return later.

In one of the interviews conducted by the Delhi-based researcher Anisha Saigal, a middle-class woman from Delhi recounts her history of negotiating gender power by hacking the paternal authority that was attempting to secure the home from globalized cable. In *Jugaad Time*, I consider domestic space-time as a feedbacked field of patterned (and controlled) but unpredictable (and agentive) intensive resonances: assemblages of discourses and practices of patriarchal control, feminist emancipation, queer-techno-pananimist-sexuality, postcolonial development, electrical "griddyness," urban proximity, and technoperceptual affordance: following Clough (2018), the diagram of an exstatic India within and against assemblages of control and accumulation. Each ecology has its own diagram, even its own method: an ecology's resonations of force, sense, and value take affective ethnography a step closer to a counteractualizing common notion, moving from an initial joyous or sad passion—connecting cables, watching American TV serials like *Santa Barbara* under the patriarchal radar—to a common notion of becoming within and against the multiplicities in force. What would a fruitful diagram here be? More specifically, how do we situate politically the changing relations of force, sense, and value that dominate, circulate, and flow in and through such practices of jugaad? Throughout this book, I define force as the set of power relations and capacities struggling to control, construct, or territorialize a given actant's relations, its assemblages, processes, and conjunctures (Latour 2005); there is also a certain, at times decisive force in the jugaadu's virtuosity. By sense, I mean the historical and embodied processes of perception, signification, sensation, tendencies, and habits that such an actant or conjuncture emerges from (Deleuze 1988a, 1988b). Value I define both in terms of monopolistic forms of capitalist accumulation, structures intensifying the exploitation of newly algorithmized and productivized bodies and relative surplus value ecologies, and in terms of the nonlinear emergence of technoperceptual "basins of attraction" in digitally networked social life in everything from styles of consuming viral memes to emergent forms of political organization such as the direct democracy and alternative currency experiments in Europe or the feminist activist movement Why Loiter?, or Dalit political organizing in Mumbai.[12] Relations of force, sense, and value limn technoperceptual assemblages through and in which capacities and affordances of carbon- and silicon-based ecologies coevolve.

Thus, as method, *Jugaad Time* makes the case for an ecological and affective analysis of a social practice that is quite simply a pragmatic and networked approach to an obstacle. The ecology and social history of this ethical pragmatism is potential and actual at once. The modes of sufficient reason employed in the practice of jugaad engage intuition as much as probability (Adorno 2013; Ansell-Pearson 2001; Bergson 1988; Deleuze 1988a, 1988b; Hardt 1999; Sohn-Rethel 1978). Raising this doubled epistemology to its ontological vitality, moving from a joyous passion (jugaad) to a common notion (workaround power/property), through the memory spores of control databases alive with what Jacques Derrida once called "archive fever," the diagram of jugaad overlaps fractally ($n-1$) with its own practice, constructing an interzone where memoir, science fiction, ethnography, and political philosophy combine to make "an affirmation of becoming." *Jugaad Time* diagrams sets of potentialities and probabilities both within and "beyond" jugaad, and, as Nietzsche urged, for the benefit of a time to come.

This is another wager of *Jugaad Time:* the diagrammatic method of affective ethnography is focused as much through acts of enabling destruc-

FIGURE I.4. Example of a life-hacking jugaad. Photo by Anisha Saigal.

tion (critique, deconstruction, parataxis) as through modes of decolonizing attention (Povinelli 2016; Strumińska-Kutra 2016; see also Hart, G.
2006, 2008). Thus the term "ethnography" is transformed and recontex-
tualized in the realm of affect; decolonizing attention affects the term
"ethnography," and the attention it takes to do this has become involved
in processes of experimentation within and against the panoptico-digital
and capitalist capture of ethnography. The diagram presented in the chapters that follow charts the actual and potential resonant fields enfolding
the practice of jugaad in forms of creativity, research, value generation,
politics, and communication. For instance, linking urban and neural plasticity to the hacking practices common in jugaad cultures yields strategic
questions for how policy is implemented, as well as how feminist, Dalit,
and queer politics is imagined and practiced. The capitalist economy in
India is actually lived by the vast majority of its inhabitants in ways that
bring to crisis the neoliberal methods of academia. Indeed, capital itself
throughout South Asia has broken with contemporary academic forms
of critiquing, arguing, explaining, or demonstrating by relating the prac-

tice of jugaad ontologically and epistemologically to both the histories of DIY cultures of everyday resistance of subaltern subjects and communities (frugal innovation), and to an as-yet-undecided, potential future in which the practice of jugaad reorients emergent technoperceptual assemblages to become resonant and functional. Each of these relations is an ethics of composing a plane of consistency. This method draws on the work of Franco Bifo Berardi, Erin Manning, Brian Massumi, and Félix Guattari. As Bifo notes, in a talk at a meeting on "Psychoanalysis and Semiotics" held in Milan in 1974, Guattari spoke about signifying semiologies and a-signifying semiotics thus:

> My opposition between despotic signifying semiologies and asignify-ing semiotics remains highly schematic. In reality, there are only mixed semiotics which belong to both in varying proportions. A signifying semiology is always shadowed by a sign machine, and, conversely, an a-signifying sign machine is always in the process of being taken over by a signifying semiology. But it is certainly useful to identify the polari-ties represented by the two, in other words the signifying semiology as a paranoid-fascist ideal, and a non-signifying semiotics as the ideal of schizo-revolutionary diagrammatization, of getting beyond the system of signs towards the plane of consistency of particle-signs. (Guattari 1984, 140)

Bifo writes, "Guattari uses here the notion of the particle-sign. This is his way of molecularizing semiosis, of seeing it as an activity of projecting psychochemical agents. We can speak of signs as material agents of semi-otic mutations. This has nothing to do with ethereal messengers of mean-ing, but instead with viral agents in Burroughs's sense who, in fact, spoke of language as a virus: signs not as pure representations, but propagating as asignifying contagion, transforming the semiotic ecosphere" (Berardi 2008, 107). Thus, what is at stake for *Jugaad Time* in the diagrammatic method is precisely a thoroughgoing displacement of representational and semiotic frameworks toward an experimental and embodied (or molecu-lar) practice of productive contagions traversing multiplicities.[13] Consider the set of jugaads operating within Prime Minister Narendra Modi's dis-course and practice of "smart cities." Through legal (security and control-oriented policy) and extralegal means (dispossession), the smart city has emerged as a form of elite reappropriation of the right to the city. Numer-ous contrasting forms of technoperceptual contagions traverse the formal and informal dichotomy at different scales: mobile phone viruses, always on sensors, habituated practices, tinkering, infrastructure-light security, data ontologies, hacking, and so on. These vectors of contagions operate through the project of smart urban (re)generation, or anti-Dalit and anti-Muslim social cleansing. An implicit and generalized jugaad governmentality makes possible both the ongoing and rapid privatization and segregation of social, material, media, and economic infrastructures, as well as the proliferation of contagious practices of commoning resources in ad hoc and makeshift pirate infrastructures and emergent technoperceptual assemblages (Jamil 2017).

In the chapters that follow, I diagram jugaad practice through its moments of corporate capture as innovation, in its dynamics of noncapitalist refusal, and in its plastic time-spaces of creativity. Thus, I hope to show how political ecologies of sensation can raise compositional questions for both radical politics and queer embodiment. What I will call the hacking ecologies of jugaad necessarily entail an ethics of experimentation. In "Machinic Orality and Virtual Ecology," Guattari locates ecologies of the virtual in practices of aesthetic and hence ethical experimentation that are "deterritorialised machinic paths capable of engendering mutant subjectivities" (1995, 90). Deconstruction of the structures and codes is a necessary but not sufficient step on such paths, and as we go further toward an epistemology of ontological common notions we are invited to take "a chaosmic plunge into the materials of sensation" (1995, 90). In India, increased inequality and unanswered injustice, rampant militarism and unchecked authoritarianism, as well as state suppression of dissent, stigmatization of nonnormative sexualities, and neoliberal policies legitimating privately owned civic "smart" spaces have all made it imperative to recast the axes of values, forces, and senses that naturalize injustice as the fundamental finalities of human relations and productive activity. Understanding jugaad's ecology of the virtual is thus just as pressing as knowing its actual ecologies of the pirate world. Following Guattari, we will attempt to move beyond the relations of actualized forces, and into virtual ecologies that will not simply attempt to preserve the endangered species of Indian cultural life but to "engender conditions for the creation and development of unprecedented formations of subjectivity that have never been seen and never felt. This is to say that generalized ecology—or ecosophy—will work as a science of ecosystems, as a bid for political regeneration, and as an ethical, aesthetic and analytic engagement" (91–92). Guattari urges us toward a fractal ontology where the subject is a fractured anchor point within "incorporeal fields of virtuality" (95). Thus, jugaad's very timely and increasingly valuable ecology of sensation can be diagrammed as a historically specific but ontologically untimely phase space of a (de)habituating body, never identical to itself, "in permanent flight on a fractal line" (95).

If a history of hacking is possible, surely its pedagogy is not. You can't teach someone to hack this or that with any degree of success without some initial, militant propulsion of something more or less anonymous and free, nonhuman, transhuman, some faceless Untimely vector. Vectors, not scales or subjects, would be the starting points of this history. Vectors are o/bjects of multiple agencies. There are, for instance, revolutionary vectors and suicidal vectors, inertial vectors and vitalizing vectors, often merged in reciprocal multicausality. Where are they tending? What's the threshold of force-contraction for vectors to act? These are the first questions in all ethical diagrams, why can't they be of history as well?

Consider revolutionary becoming a vector. If we define a vector field of politics as both virtual and actual at once—and we don't mean by this now superseded word, virtual, something having to do only with inanimate technology, or this yet to come word, actual, with what probable futures we can affirm today—a vector field of politics is virtual and actual at once because in it durations, affordances, and prehensions are constantly transforming the conditions of becoming, even if only at the molecular, and strictly speaking nonhuman level. Hegel said that quality is nothing other than a contraction in quantity. Scales are kinds of contractions in quantity. Today, scale is a habit of thought. Vector is a practical reason. Enfolded in that (un)reason, revolutionary becoming is a militant refusal of the timely and "actualist" politics of identity that characterizes our contemporary biostate of exception.

There is a better way to remember: not the history of trauma or the affirmation of a braying ass, negation and resistant-becoming, but rather mutations in experimental vector fields. What were the vectors? How did they intersect? With what feedbacked subjectivity and interactive technology, to take only two contractions of difference? The vector character of all becomings is especially important to mark when thinking about the commoning pedagogy of hacking. Hackers are always, to however minimal a degree—and we must speak here, at the start, of degrees, intensities, constrictions, and elasticities—teaching us to place ourselves outside the Law. Which Law? The Law of Habit. The Law that Hackers break ontologically is the Law of Habit. In time of revolutionary becoming, hacking becomes a virus *dissegenating* throughout an ecology.

I.

When the phones stopped ringing, our bodies, our brains were still vibrating, still moving along a habitual vector that had suddenly been separated from what it can do, triggering a kind of collective gnashing of teeth. Headaches were initially reported along an arc of Western Europe and the North of the UK; linked to the singular role of the jaw and throat muscles in the enunciation of the diphthong associated with "o" in most of the languages across this region, and coevolving with the mobile phone chip and its signal. On average, and controlling for other variables, the headaches lasted for two years. In some subjects each episode lasted days rapidly, moving from hemisphere to hemisphere, in some the habituated hunger for a techno fix successfully sublimated through the InterFace. Many preferred suicide. This is not their story.

II.

How did digital code coevolve with the mining of Columbite-tantalite, or coltan? This is immediately both a political and aesthetic question. A vector often synthesizes ontologically diverse phyla: biopower brought together the biologized, hence ecologized body with the order of racist discourse; media brought together perception and matter enlivening both through prehensive mutual causality. In the first years of the Reinvention, biostate spun off different studies in creative clusters across the galaxies, cobranding a process for a scientific and total geology of practices, but all they yielded were techno-fossils, fragments, indications, abysses, but also certain vectors if analyzed correctly. An unprecedented proliferation of regimes of truth consolidated digital borders, little identikits of paranoid conspiracies, violent archive fevers raised to a harrowing pitch, and voracious attention vacuums. The faces emerged, dissolved, with each new financial crisis, war, fashion season, and Holidays. They were this month's Face, or that month's Face, a regularity of rhythmic loops, duly forwarded in the endless workflow of Xdata. The Untimely Militant collectives that have overlapped with for instance Maroonage, Insurrection, Revolt, General Strike, People's War, militant study groups, and community and union organizing always constitutes a certain zone of action and indetermination from its own refusal of privatized value. They discovered then the stochastic resonance between a chemical and geological phylum forming parallel evolutionary trajectories of carbon- and silicon-based life. Today, we know to ride that multiphylum like surfers used to ride waves, chance integrated into "feedback" devices which are "the events of History; and that it is instantly incorporated by the whole so that it appears to everyone as a manifestation of providence, etc. But this misses the point. It is not a matter even of showing that such syntheses are possible, but of proving that they are necessary: not any particular one, but in general that the scientist must adopt, in every case and at every level, a totalising attitude towards his subject matter" (Sartre 2004, 16). Anything that irreversibly transmits flows across vector gradients can be a feedback device: the abject cyborg in biostatecraft as much as the hacker's virtuosic assemblage/practice. I think we should generalize this term to better think the polyvalent, or nonlinear history of vector fields and their ethics, or politics. But at that time, long before the series of seemingly unrelated ecological phase transitions that came to be known, perhaps ironically, perhaps strategically, as the Minor Events that kicked off first in Bhopal, we could source coltan from our global mining partners, sometimes from Brazil, sometimes the Congo, little concerned with bloody civil wars, enforced poverty and sex trafficking, environmental racism, privatized detention centers, the widespread use of enslaved child labor, and outlawed labor unions. Over a period of some centuries, the insurgencies against oligopoly and settler occupations became too massive, too often recalcitrant, and so without a thought for its effects, this ecocidal exploitation of land and labor was righteously proclaimed "unsustainable," in time putting accumulation pressures on the extraction machines, and in turn threatening the production of neuro-electronics and hence affecting the sustainability of the capitalist evolution of binary code. Crisis after crisis, disasters becoming everyday, toxicity become species-endangering—these and other "innovative" or "creative" combinations of value, sensation, matter, energy, conviviality, and information were sought, produced, curated, even. Meanwhile, here and there, sometimes without hope of communicating their energies, people of the phase transitions gathered together the resources of a digital and material Undercommons. This is when the jugaadus got involved again, inaugurating an era of becoming within and against the Minor Events.

III.

A mobile phone used to emit a kind of pulse, a nonradioactive electromagnetic sensing pulse.[1] It was, to use an old phrase, a probe head. The Minor Events took that pulse and transmuted it by reengineering both its energy (speed, flux, acceleration, heat: its vector field) and its information in new forms of "autonomous" assemblages. They dubbed it the Reinvention: Re-vector Your Life!™ (Part of the Sourcestyle Sub-brand ecology). Linking the world's emergent artificial intelligence to a massively thickening carbon and silicon based neural net, this pulse of energy and these packets of information (data reinventing life), incrementally and indiscernibly recoded the operating systems of millions and then billions of "autonomous" handheld, habitually connected devices, in mutating technoperceptual assemblages. The Series of Minor Events triggered a global conflagration between what was already by then postcapitalist security and the disparate but global Pirate Kingdoms circulating through their war machines. The Collectives for Revolutionary Jugaadus started then—each with their own solidarities, some dovetailing with control regimes, some or all failing in their own time, yet each expressing and composing an ineffable and non-human capacity for collective, if Untimely, emancipation. These collectives for a radical practice of hacking into the walled confines of capitalist code proceeded at first with little hope of success. After scale, Hope was the habit the Minor Events discarded next. But each time and in unlikely locations, exiting the monopoly rent regimes of debt, security, competition, and IP, the vector diagram was first affirmed in a monstrous and collective revolutionary becoming that could have no recognized name or known historical value—only experimental organizational practices in continuous reciprocal causation, counter actualizing their vectors to become—and later controlled as a lifestyle brand called Reinvention.

IV.

Objects become ecologies in hacking diagrams, shifting attention to what, say, a mobile phone in Bhopal can do. Jugaad diagrams can become "minor events." Bottlenecks, piracy, commoning, and sabotage in the logistics of global value ecologies were vectors recomposing value. New mechanisms emerged for their capture, growing stronger in proportion to their violence. In Hindutva-administered Bhopal, the Biostate University campuses were shut, condemned for becoming "hotbeds" of revolutionary commoning. Chancellors denounced the movement's "naive susceptibility to communal manipulation by the neo-Naxalite, Dalit-led infestations from Muslim Old Bhopal."

The soil beneath these campuses was enfolding hidden geological processes of crystallized methyl isocyanate from the Free Factory Zone reacting with groundwater, the remaining lake, and through sewage tunnels, seeping into the underground passages dug out for the rebranded "kabad to jugaad" (junk to workaround) microfactories. Meters below the empty showrooms of MP Nagar, migrant Dalit workers became virtuosic jugaadus, carving out a vast maze of microfactories, daily breathing in the lachrymose miasma of cyanate vapors. A chemical phylum, poisoned, debilitated, and mutating, primed some areas of the city and their populations for carbon- and silicon-based phase transitions.

How did the new revolutionary collectives diagram the changing ecology of their own becomings? Which intersections of gender, caste, race, class, religion, sexuality, or ability would be affirmed in their practices was a matter of historical transvaluations. Biostate's administration of caste and identity more generally "reinvented" value extraction as distributions of an invented, but probabilistic population's "affective" (in)capacities. This is what Reinvention helped to automate, financialize, and market. Radical collectives experimented with fugitive organizations of politics, techno-aesthetics, and value production, expressing old and new relations of a revolutionary Undercommons within and against Hindutva Bhopal. Reinvention co-opted this countermemory itself as a source of monopoly rents for "Toxic Wasteland Tourism": rebrand a Bhopali identity, monopolize its speculative value ecology, develop gradients of subalternity in populations whose incapacities become dynamic control parameters, or affects. It was a moment of transition, full of specific morbidities. The value capture mechanisms of what would become the Reinvention brand were assembled simultaneously through the circulation of a wildly successful "Bhopal's Minor Events of Revolutionary Jugaad" series, and through the algorithmic and speculative aggregation of the "affects" of those memes. Speculative measures of jugaad practices also emerged. Queer Undercommons, diagramming a nontotalizable, transvaluating curiosity across species, geology, energy, and information, heralded a brand new social synthesis. Leaders emerged: their becoming discernible was a trap.

THE AFFECT OF JUGAAD
"Frugal Innovation" and the Workaround
Ecologies of Postcolonial Practice

The event in the strong sense of the word is therefore always a surprise, something which takes possession of us in an unforeseen manner, without warning, and which brings us towards an unanticipated future. —F. DASTUR, *Phenomenology of the Event*

Jugaad innovators . . . constantly employ flexible thinking and action in response to the seemingly insurmountable problems they face in their economies: they are constantly experimenting and improvising solutions for the obstacles they face, and adapting their strategies to new contingencies as they arise. . . . The sheer diversity, volatility, and unpredictability of economic life in emerging markets demands flexibility on the part of jugaad innovators. It demands that they think outside of the box, experiment, and improvise: they must either adapt or die. —N. RADJOU, J. PRABHU, and S. AHUJA, *Jugaad Innovation*

Introduction: Improvisation and Contingency,
or the Affect of Jugaad

This chapter engages the ongoing conversation around piracy in postcolonial media studies and social geography across three areas. First, focusing on the ecological processes of the social practice of jugaad (workaround), I show how its strategic deployment, production of time-spaces, and digital media assemblages habituate heterogeneous populations in India toward innovation. Developing work in postphenomenology and nonrepresentational analyses, I draw out what researchers have called the technicity of affect (following Ash 2010, 2013; Clough 2010, 2018; Heidegger 1962, 1977; Mackenzie 2001; Massumi 2002, 2011, 2015a; Stiegler 1998; Thrift 2005, 2006). The social practice of jugaad allows human-technical assemblages to intervene specifically in the material contexts of (in)subordination through various forms of technology. Second, to understand both the event and the ecology of jugaad, I turn to an important concept in postcolonial criticism, that of translation, to understand how the affects of jugaad are translated as both habit and its modulation. The relevance of postcolonial media studies for postphenomenal affect studies will also be argued for throughout this chapter through a consideration of the materiality of subaltern agency itself, translated through an "ethological" or ecological frame (Ash 2013, ; Deleuze and Guattari 1987; Manning 2013). Human agency is shown to be distributed across material, technical, and intensive objects and processes (Shaviro 2014). Finally, by developing strategic feedbacks between postcolonial media studies and affect studies, I draw out the political and methodological implications of this analysis for nonrepresentational and postphenomenological methods of affective ethnography.

This chapter articulates affect with the everyday practice of jugaad, or frugal innovation, in India. From a marginal practice of subaltern communities (and marginal to the normal legal subject before the law) jugaad has become an important affective atmosphere ("a term that refers to the circulation of perturbations to produce space times local to technical objects" [Ash 2013, 20; Michels and Steyaert 2017) in India's postliberalization sensorium [i.e., roughly post-1991]). Its extralegal connotations translated into "disruptive innovation," jugaad is enthusiastically celebrated as frugal creativity in contemporary Indian management and marketing discourses and practices, as well as across the representational strategies of "digital cool" circulated in the old and new media (Mankekar 2015; Nayar 2012; Sundaram 2009). This capture aims to incorporate jugaad into the intensification of work in neoliberal India: use what you have to innovate; don't rely on patronage from the Ma-Baap (parental) state (Guha 1983); take inspiration from the entrepreneurial aura of the *acche din* (good days) of an India Shining (for upper caste Hindus) (Boddy et al. 2015).[1] In everyday practice, jugaad is performed when conditions of work or life come up against obstacles. In this sense, the affect of jugaad is the capacity to move from a state of relative inaction or blockage to an improvisational situation.

While the reigning popular discourse on jugaad is a moral-individual one (see Kumar 2009), in which the ideology of agency in the jugaad lies firmly in the moral decisions of the individualized, impoverished, debilitated, and/or frugal tinkerer, this chapter draws out the distributed nature of jugaad across its associated contexts and shifting human and nonhuman time-spaces. As such, jugaad is understood as an event in the affective processes of an ecological assemblage of carbon- and silicon-based life. The importance of jugaad as a focus for postphenomenal geography and affective ethnography lies in the way the practice articulates the interstices between elite and subaltern life, postcolonial studies and affect studies, subaltern workaround cultures and neoliberal strategies of frugal innovation. Indeed, the term jugaad comes out of subaltern, or "nonelite," strategies of negotiating conditions characterized by extreme poverty, discrimination, and violence, which, rather than competing or winning, are experiments in getting over the next hurdle confronting socially and economically disadvantaged communities. In this chapter, I elaborate forms of jugaad practice that Elizabeth Povinelli, following Michel Foucault, calls "subjugated knowledges" which are "mixtures of rituals and makeshifts (bricolages), manipulations of spaces, operators of networks" (Benjamin 2011; Chatterjee 1995; de Certeau 1984, xiv–xvi; Foucault 2003, 2008; Gupta 2012; Marx 2010; Pandey 1990; Povinelli 2016; Prakash 1990); the event of a jugaad emerges from social microcosms that are structured by intersecting wills to power, properties, relations, and processes in which actions are governed by often nonstandard logic and organized according to practical life without having any official rule boundaries (Bourdieu 1996, 226–227). The affect of jugaad translates power differentials across capitalist innovation and subaltern practices of everyday life in India today (Cohn 1987; Guha 1983; Larkin 2008; Scott 1985). Focusing on the deployment of jugaad in both mobile phone marketing and everyday practice, this chapter argues that the "controlled contingency" of marketing discourse is constantly being exceeded by the changing capacities and habituations of mobile phone users. However, contemporary business discourse positions the practice of jugaad as what innovation must become in the new global economy after the 2008 financial downturn. For instance, Navi Radjou and colleagues, in their popular management book *Jugaad Innovation: Think Frugal, be Flexible, Generate Breakthrough Growth* (2012), define "jugaad" for entrepreneurs as constituted by six basic principles: seek opportunity in adversity, do more with less, think and act flexibly, keep it simple, include the margins, and follow your heart (19–20). These management academics and social entrepreneurs offer up jugaad practices to drive a firm's resilience, frugality, adaptability, simplicity, inclusivity, empathy, and passion, "all of which are essential to compete and win a complex world" (20).[2] While business scholars have deployed jugaad as an image of competitive advantage in a time of austerity, multiple and often surprising systems of value emerge from the practice of jugaad in everyday life. The intersection between the macro business rhetoric that claims jugaad as innovation and the everyday practice of subaltern consumers is analyzed in this chapter to understand affect as spatial and economic, as well as embedded in ordinary experiences of "Indian" frugality, precarity, and agency.

To understand this more concretely, let us briefly consider an event of jugaad performed by Haier, a Chinese consumer goods company, cited by Radjou and others (2012; see also Chen et al. 2013). When the company learned about the clogged drainpipes in its washing machines, purchased by rural customers who were using the machines to wash vegetables (a widespread jugaad across rural China), employees began innovating devices with wider pipes that could handle washing vegetables, and eventually further modifying the machines so that they could peel them as well. Such innovation-oriented customer service has given Haier a larger market share in the fast-growing consumer goods market in China.

In this example, the everyday practice of repurposing the capacities of the washing machine is translated into the corporate condition of technological innovation and adaptability to contingency. In a complex feedback that constantly displaces the dichotomy between intellectual and manual labor, the farmer's habitual jugaad eventually produces new habits through newly innovated technologies, which in turn proliferates a culture of jugaad in the firm. For the firm, then, the circulation of jugaad as an affective disposition or atmosphere of being primed as "nimble-minded and nimble-footed" in the context of "emerging markets, which are characterized by extreme unpredictability," allows for an ongoing recalibration of the consumer sensorium through innovation (Radjou et al. 2012, 94; Adorno 2013; Ash 2013; Harrison 2000; Massumi 2002; Michels and Steyaert 2017). In this example, the literal blockage of the drainpipes is cleared up through specific kinds of jugaad at different scales of practice (farmer, firm, market share) and through different technologies (washing machine, drainpipes, modified drum) and materials (vegetables, soil, water). These different scales, materials, and technologies are involved in translating and synthesizing unequal relations of power and structures of exploitation in contexts also characterized by extreme poverty (Asad 1975; Bassnett and Trivedi 1999; Berardi 2009a; Bhabha 1984; Cheyfitz 1991; Haraway 1987; Niranjana 1992). The subaltern farmer and the value-adding firm are related through different forms of power, accumulation, interests, and value; as Radjou and colleagues (2012, 94) note, most firms would declare the warranty, and so any access to customer service would be void due to the misuse of the machine, whereas for the farmer economizing on time, water, and labor the jugaad of repurposing the machine frees up scarce resources, one of which is time itself. If, recalling my definition of affect as embodied passage, a materialist diagram of such jugaad events will follow the durations that gradually stabilize in its wake, it is in translating and synthesizing the extralegal jugaad into a viable and scalable value proposition through innovating new machines that the firm is able to extract both more surplus value from the machines and secure new consumer segments.

Or consider another example of jugaad in practice, the relation between mobile phones and the language of "missed calls" that has emerged in many parts of India. Since around 40 percent of subscribers have no more than Rs. 5 on their phone, and without resources to "top up," a new jugaad emerged: give a missed call to communicate with an interlocutor. Costing nothing (following the shift to the "calling party pays" system of accounting in the early 2000s, the caller would hang up before the call is answered), the missed call communicates through the network despite the caller's lack of resources. This is a simple example of the emergence of a jugaad; from the obstacle of a phone without credit, a form of communication remains functional. Unlike in the example of the Haier washing machines, here there is nothing specifically extralegal in the jugaad, but a new workaround develops, bypassing the limitations of resources.

Following Kevin Hart, John-David Dewsbury and James Ash, whose researches into contingency and event have had an important impact on recent work in human geography that theorizes how affect is constitutive of and woven into everyday life (Anderson and Wylie 2009; Ash 2010; Dewsbury 2000, 2003; Harrison 2000; Massumi 2011; McCormack 2007), this chapter aims to emphasize the "more than human" phenomenality of jugaad by shifting attention away from jugaad as a discrete object or event that appears for human consciousness toward how and with what affects jugaad emerges ecologically (Hart, K. 2007, 39). I ask, what are the processes by which jugaad events come to be both "potentially intelligible for consciousness" (Ash 2012, 188) and a potentialization of jugaad's ecology itself? Throughout, I refer to affect in the Deleuzian sense: as the durational passage from one state to another in an encounter between two or more bodies (human or nonhuman, organic or inorganic), which either increases or decreases a body's capacity for action (Deleuze 1988b; also see Ash 2013, 188; Clough and Halley 2007; DeLanda 2002, 62; Manning 2013; Massumi 2002, 2011; Rai 2009). Affect is the bodily capacity to sense and act; more a transitional duration than a fixed state, affect is the event-potential activated in and through the passage from one state to another. As such, affect involves both contingency and its history in any given context of interactivity between humans, technology, and biopolitical life (Ash 2010, 2012; Thrift 2004).

The analysis of affective geographies has helped to reframe the bodily experiences of boredom (Anderson 2004, 2005) and (dis)comfort (Bissell 2008, 2009) and skilled practices such as gaming, art, mobile phones, and busking (Ash 2010, 2012, 2013; Bogost 2006; Massumi 2002, 2011; Simpson 2008, 2009). However, as Ash notes within geography much less work has been conducted on the ways in which affect can be actively manipulated for commercial and economic ends in the design and production of consumer services and goods (see Ash 2010, 654). By taking the intentional and personal subject out of phenomenality, Ash's work focuses on the affective encounters between humans and technologies in order to develop an understanding of technology as generative of inorganically organized affect, which enters into and reorganizes the affective thresholds of the body (Roelvink and Zolkos 2015; Dale and Latham 2015; Duberley et al. 2017; Duff 2010; Gascoigne et al. 2015). It is through such "turbulent" affective processes that the spatiotemporal dimensions of sense are organized. In this view, phenomenality encompasses the past, present, and future as specific modes of potential and how "these modes are actively fixed for human perception as a kind of spatiotemporal envelope through a variety of body–technology assemblages" (Ash 2012, 188; Ash 2013). In short, the phenomenality of jugaad shifts *attention* to the ways in which these processes operate and become habitual in local socialities and contexts.

Differentiating itself from most postcolonial media studies that continue to focus subaltern agency on the intentional subject, this chapter attempts a renewed understanding of postcolonial translation as distributed and ecological processes of converting movement, contingency, and matter into specific kinds of value.

This chapter draws on ethnographic work on mobile phone ecologies in Mumbai and Delhi between 2009 and 2017. In interviews with mobile value-added services (MVAS) executives, sim card vendors, mobile phone repair workers, everyday mobile phone users, and media sociologists, the word "jugaad" came up repeatedly to describe both what the human–mobile phone assemblage could do, and what one needed to do to get the most out of that assemblage (that is, to get over an obstacle, get the thing moving). The problem that is posed in the increasing acceptance of jugaad as an everyday practice of nonelites in India *and* as a business model for global firms is explored in this study through methods drawn from discourse analysis, affective ethnography, and mapping ecologies (processes of emergent properties, and ecologies that mutate through creative evolution—Bergson 2012; DeLanda 2002, 2010; Warf 1990).

While the focus in this chapter is on the volatile context of the Indian mobile phone ecology, the broader implications for a radical diagrammatics of affect will be explored in this and subsequent chapters. If mobile phone practices are enabled by specific affective environments, the embodiment of this space in different forms of affect, or capacity, is both crucial to its marketing and its mutations. To think through the affective dispositions of play in jugaad with the assemblage of mobile media sweeping across India today, I focus on its emergence as a habituated sensory-motor circuit of mobile digitality (Ash 2010; Massumi 2011, 89; McCormack 2005). Through a jugaad, a place of obstacles to work, value, or desire becomes a relatively potentialized, improvisational, and contingent space through the mobilization of whatever resources are to hand. The argument is that jugaad ecologies affect a passing from stasis (obstacle, interdiction, hurdle, blockage, problem) to improvisation (passage, event, contingency, value, movement), and translating new or added value from this affective passage is an aim of contemporary mobile phone business practices drawing on jugaad.

I proceed by analyzing both the "controlled contingency" and "ecological emergence" that form the antagonistic heart of the postliberalization uptake of jugaad in India's mobile ecology. In doing so, I begin by considering how the business model of the mobile phone provider Bharti Airtel has become an example of frugal innovation, or jugaad. I argue that throughout the Indian mobile ecology, phone companies have not only successfully adapted jugaad to their business model, but also adapted customers and employees to a new sensorium through a specific spatializing and temporalizing image of jugaad. I then consider specific Airtel ads that appeared on Indian TV in 2010–2011, and show how the image of jugaad is figured in the contingency of mobile phone practices and in the marketing communications of mobile phone providers. The conclusion considers what is at stake in this analysis by limning the limits and possibilities in the method of understanding the affect of jugaad as ecology and event at the intersection of space (the mobile phone ecology) and emerging markets (macro and micro) in India.

India's Mobile Phone Ecology and Airtel's Jugaad

As others have noted, telephony in the Indian subcontinent has a long history, thanks to the communicative needs of British colonialism. For example, the first telephone service in the subcontinent was introduced in 1881 by the colonial government in Kolkata.[3] Following independence in 1947, the state monopoly on telecommunications continued for fifty years, limiting access to communication technologies to most Indians. The postcolonial state administered media ecologies for specific developmental goals and in alliance with ideologies of upper-caste nationalist hegemony. Subaltern communication practices, however, continued in the jugaad time of hacking ecologies.

After liberalization of the Indian economy in the early 1990s, the telecom industry was marked for rapid reform and by 1994 a process of privatization was initiated. Rapidly, "frugal, disruptive innovation" (Radjou et al. 2012, 100–103) became the industry's hallmark. As we shall see, Bharti Airtel's own disruptive innovation and self-positioning as an "aspirational and lifestyle" brand helped create the organizational strategy of the asset light telecommunications firm in India, as well as marketing communication processes associating Airtel with "digital cool." Despite the tendencies of transnational capital toward oligopoly, corruption, and elite hegemony, today forms of piracy throughout South Asia and the rest of the global South have both deepened the habituation of populations for copyrighted materials, software, and forms of prosumption, and created the conditions in which media mutations in assemblages of silicon- and carbon-based life push the limits of what Ash calls material and intensive thresholds (2015a, 2015b). For Ash, technical objects have a relative autonomy based on their material and intensive thresholds; while humans design thresholds into objects to attempt to control the kinds of affects that are generated, "technical objects always have the potential to exceed the intentions of their design because they have a homeostatic autonomy that exists outside of any one human's grasp. Material components and thresholds can be reworked, modified or simply broken down, which in turn generates a whole new set of thresholds within which affects can operate" (Ash 2015b, 12). In India, the mobile phone market is traversed by multiple competing attempts to organize the material thresholds of mobile devices. As Ash argues, taking the notion of the inorganic organized nature of technical objects seriously allows us to focus on how affects are generated through encounters between technical objects within an ensemble and translated, which in turn generates particular affects as humans encounter them (2015b, 12–13). Thus, while there are more than one billion mobile handsets in India, and while the national market continues to be the most competitive mobile service market in the world, innovations in handset design focusing on battery life and water resistance, the introduction of 4G data networks, dual sim-card phones, and shifting temporalizing and spatializing social practices all bear the marks of struggles around the organization of the mobile's material and intensive thresholds and the particular affects that users encounter through them.

Smartphones, the cheapest of which (the Magicon M3 Atom) sells in India for about Rs. 6000 ($90 USD), allow for the rapid, on-demand consumption of media such as ringtones, wallpaper, videos, TV episodes, and film songs. Much cheaper handsets are widely available, and new, even cheaper models enter the market regularly (Radjou et al. 2012, 100). Significant for this fast-changing economic sector are the ever-increasing MVAS available in India, such as media delivery, banking notifications, astrological predictions, weather and traffic updates, heterosexual soft- and hardcore porn, news, jokes, and latest offers from a variety of legal and extralegal businesses. These habituated patterns of consumption become potential sources of new productivity and new products as MVAS management information systems (MIS) reports correlate fresh data with patterns of use in other datasets (Massumi 2011, 2015a). Indeed, matching patterns of interaction is one source of value creation and capture in this ecology.

I first encountered the widespread use of the expression "jugaad" as I began interviewing and socializing with MVAS executives at an up-and-coming firm in Delhi-Noida. The firm offered products and services across the range of MVAS, which at that time was just being developed in India. Products such as entertainment content, contest management, news and information dissemination, social marketing campaigns, supply chains, and logistical uses of mobile phones were all handled by this company, in differentially scaled projects pursued by competing teams. The management executives were as focused on the habituation of populations of mobile phone users as they were on maintaining a strategic, that is, competitive, relation to risk.

> *Fieldnotes Excerpt*: Sitting in the second-floor offices of Mobile Value in Noida. Noisy, dusty, lawless Uttar Pradesh (dangerous to be here after dark because of the corrupt police). Outside the floor-to-ceiling windows, the city is bustling with life below, busy, horns, traffic, congestion. I'm sitting in the over-air-conditioned second-floor conference room surrounded by clear glass walls; across the hall on a white wall are posterized images of John Lennon, Jim Morrison, and Bono. Everyone is excited about the scope of MVAS in India. Everyone in the industry at least. The other night, the second all-nighter in a row—definitely too much for me, horrible skin thing on my nose of all places!—speaking with Gita's friend, Ambar, he was very excited about the innovations and possibilities for VAS, and the scope of mobile telephony more generally. Highlights were: in 2003–2004 Reliance mobile revolutionized the industry by bringing costs down to about R. 1,200/mo. Ninety-five percent of customers are prepaid cardholders. The pattern of communication for most people is to receive calls, which is free. Most people do not own a handset for more than twelve months. The aim of MVAS is perceived value—what I have been calling self-expression products: callback tunes, wallpapers, and the like. The biggest sense I got from him was that things are changing rapidly: there is a lot of innovation happening in the industry, and the regulations are coming down. Lots of scope.

The affects of jugaad circulating in the MVAS industry were both part of the capacities of the mobile phone and its technical infrastructures and more generally about the new capacities of capitalist innovation in the postliberalization economy. Jugaad expresses a strategic multiplicity. In one sense there is a shift from a state of blockage or obstacle through the whirlwind of contingency and intensity to a state of increased value or power. Jugaad allows for the harnessing of the ecological powers of the embodied mind, its affordances, tendencies, and capacities for improvisation. In another sense, for the professionals who inculcated a kind of culture of jugaad in the firm, this was also a way of invoking a connection to the working-class or lower middle-class backgrounds from which many of them came, a version of "keeping it real." Here, the affect of jugaad is doubled by the image of postpolitical chic, where the identity of the non-elites gives added grit and reality to a professional still able to pull off a difficult project within the firm. This culture of jugaad had become a style of management through permanent intrigue and competition: Who had pulled off which jugaad, and how has it strengthened the competitive advantage of the firm?

Below, I analyze the blurring of place boundaries effected through mobile consumption by looking closely at the style of consumption being offered by the major mobile service providers in their various advertising campaigns, and contrasting that with recent consumer activism contesting the fine print in these ads. To contextualize the Indian mobile phone ecology, I conclude this section with a twined case study of India's largest mobile network operator, Bharti Airtel, and how in North Indian villages mobile phone users are consuming mobile media in a "battery revolution."

Blurring Boundaries:
The Impact of Mobile Phone Consumption in India

Airtel, formerly known as Bharti Tele-Ventures Limited, is today India's largest mobile-network operator (MNO) with nearly 280 million customers across its operations (September 2013); operating in twenty countries across Asia and Africa, with headquarters in New Delhi, the company ranks among the top four mobile service providers globally in terms of subscribers. It offers a range of telecommunication services, from fixed lines, to 2G, 3G, and 4G wireless and high-speed DSL broadband services, mobile commerce, Internet Protocol television, and end-to-end data and enterprise services to international bandwidth access through gateways and landing stations (Aggarwal and Gupta 2009, 108; Pousttchi and Hufenbach 2011).[4]

As recent commentators have noted, the mobile market in India is characterized by intense competition dynamics, shaped by liberalization, convergence, and new technical standards (Pousttchi and Hufenbach 2011). Increasingly, firms from other areas of the economy compete with MNOs for difficult-to-find brand-loyal subscribers. As noted above, the strategic focus has shifted from simple network access to MVAS, and increasingly from MNO-generated products and services to the new "value propositions" of nontelecommunication companies, such as established hardware manufacturers or IT companies (Pousttchi and Hufenbach 2011, 299). It is in this highly volatile context that Bharti Airtel, ranked sixth in India's top fifty brands (according to Brandirectory.com), has been heralded as the epitome of a jugaad-oriented firm. Radjou and colleagues connect jugaad practices to Airtel through its "frugal strategy" of remaining asset light. In the early 2000s, as the mobile revolution was taking off in India, Airtel was short of both the capital and the technology that it needed to scale up its business. "Today, IBM manages Airtel's IT infrastructure while Ericsson and Nokia Siemens Network manage its network infrastructure. . . . By transforming fixed technology costs into variable costs, Airtel not only succeeded in getting more for less, it also did so at breakneck speed—at times signing up as many as ten million subscribers per month" (2012, 62–63).

For Radjou and colleagues, Airtel's particular jugaad lay in the innovations in organization, infrastructure, and outsourcing initiated by its founder, Sunil Bharti Mittal, to create a relatively asset-free telecom service provider. Ironically, what this implies is that Mittal was able to develop Airtel through a low-risk, efficient organizational structure, thereby minimizing the impact of contingency. Airtel's jugaad is that it has managed to carve out a dominant position in the telecom industry in India, and increasingly in other parts of the world by reducing the risk of contingency through infrastructure outsourcing (Radjou et al. 2012). Transforming the value chain through interfirm collaboration, Airtel was one of the first non-Western organizations to outsource its infrastructure and, in the same stroke, compose itself with the supply chains of global telecommunications oligopolies (Warf 2007).

In this organizational innovation and strategic business context, Airtel's jugaad enabled a competitive advantage in the telecom sector. This is a far cry from the average Airtel subscriber in India, over 40 percent of whom at any given time have less than R. 5 (£0.05) on their phone, but who have developed various kinds of jugaad for communication, the most well known of which (as I noted) is through the missed-calls system. Ravish Kumar (2013) points to other everyday mobile jugaads, specifically allowing us to see the emergent phenomenality of jugaad through its human and nonhuman ecologies:

Atul told me that a twelve-volt battery is allowing us to run our TV and watch prime-time shows because there is no electricity in our village [गाँव में बिजली नहीं आती]. When I asked him how, Atul told me that they had installed separately a fifty rupee "plate" inside Dish TV's set-top box [हमने डिश टीवी के सेट टाप बाक्स में पचास रुपये में अलग से एक प्लेट जुड़वा ली है]. They then connect a twelve-volt battery to the plate, and the conversion from DC to AC keeps a black and white TV running. The battery is a "life line" [लाइफ़ लाइन है]. . . . Most people have these batteries. Every morning the village folk give five rupees to charge them through a generator. Through that battery then people run a light bulb, a mobile phone, and a laptop. . . . He was saying that a new Chinese convertor/charger is coming out that can convert the direct current into AC and so charge a mobile phone. What then do you do with the phone? You can get twenty films on a 2-gb memory chip. Everyone pays thirty rupees to download twenty films [बीस फ़िल्म डाउनलोड करा लाते है सब]. *Bullet Raja* [a popular Bollywood film] was released on Friday, and on Saturday a high-definition version was available for internet download. Now any shopkeeper in the village can go to Bareilly (the closest urban area), download converted films, and bring it back to the village. For fifty rupees. He charges two rupees per file transfer to thousands of phones. Putting in fifty rupees, he has made between one thousand and two thousand rupees. Through a twelve-volt battery the villagers charge their mobiles and in the evening see *Bullet Raja*. The DVD player is useless. There's no electricity, right? Everyone's watching TV on a fully charged mobile. Atulji started telling me about one more thing, that the shared laptop is not used for surfing the web. Everyone is watching films and porn [सब फिल्म और पोर्न देख रहे है]. The internet is on the phone [इंटरनेट तो फ़ोन में है]. The laptops are still selling, but another trade has started [एक और धंधा चल पड़ा है]: They "window corrupt" phones. To fix the window corruption a shopkeeper will charge three hundred rupees. The shopkeepers turn even small phone problems [मामूली प्राब्लम] into a case of windows corrupt. (My translation)

This report, written in a breathless Hindi prose echoing of rumor, is significant in thinking through together questions of postcolonial translation and the phenomenality of jugaad. First, we see that the mobile jugaad is rooted in specific kinds of transnational media habits formed over varying timescales and tied to the gray and black economies of the global South: Hindi TV and film, Western pornography, Chinese mobile phones. These habits have their own material and intensive thresholds; for instance, the passing reference to high-definition digital copies of films indicates an important criterion for downloaders, and the materiality of being off the energy grid has its own thresholds of connectivity, as generators and batteries extend the uses of laptops, TVs, and mobile phones. Second, we see how this emergent digital jugaad economy intertwines both the infrastructures of subaltern media assemblages and capitalist value extraction. Examples are the shopkeeper who has access to high-resolution film files and can generate exponential wealth through the illegal download business, and the windows-corrupt *dhandha* (a Hindi word that translates as both "business" and "sex work") that plays on fears of computer viruses and manipulative merchants. Finally, returning to Ash's conception of material and intensive thresholds, we see in this example how material components can be reworked or modified in an extralegal jugaad (a plate grafted on to a set-top box, DC into AC, etc.), and from another direction how the jugaad of the corrupt windows generates a whole new set of thresholds within which affects operate. Note then in terms of the postcolonial translation of scale in practices of jugaad, and the shift from a pragmatic to a strategic context: at the level of subaltern, or nonelite practice, jugaad is the affect of moving from obstacle to improvisation, while at the level of the mobile phone network provider, the jugaad is about innovating to achieve brand dominance across the market. Not surprisingly, this (non)translation of scale, its differential relations of power and value extraction, is covered over in the celebratory discourse of Radjou and others.

Jugaad as Event and Ecology for Sensory-Motor Habit

If we turn to Airtel's televised marketing campaign in recent years, we see more specifically how the organization's branding of jugaad as an everyday mobile phone practice is being enfolded into a broader cultural or even behavioral experience of value in mobile ecologies. In the following analysis I wish to connect a marketing image (digital cool) with the processes that it attempts to capture and revalue. Following Nayar (2012), I define digital cool in terms of processes that through binary code and "total" connectivity produce a relatively high and dense degree of interactivity, and where "there is extensive user-generated content and multilateral (one to one, one to many) communications" (Nayar 2012, 3).

The associations of the mobile are not merely the discernible objects and distinct parts of a closed system as contemporary mobile marketing attempts to predict it. The phenomenality of jugaad puts into focus, first, the coevolving bodily capacity to affect and be affected, and, second, the increasingly dynamic and stochastic nature of the mobile-human connectivity passing through thresholds of density, noise, and repurposing, and, third, new practices of "commoning" digital information, open-source code, and subaltern patterns of interaction (e.g., open-source cocreation, virtual private networks [VPN] to peer-to-peer networking, hacking and cracking phones; see Rossiter et al. 2012; Wark 2012). Such associations open the mobile media assemblage to a variety of material, work, semiotic, psychic, perceptual, and informatic translations as well as new forms of politics and political subjectivity (Ansell-Pearson 2001, 40–41; Bergson 2012; Guattari 1995; Nayar 2012).

This returns us to our initial discussion developing a feedback between postcolonial translation and the phenomenality of objects in the social geography of affect. The Airtel advertisement analyzed here assimilates the differences of jugaad—as an extralegal, improvised, subaltern practice that forms assemblages of technological and human movement—through the translation of the sensory-motor habits of mediatized images and through designing encounters between humans and technical objects within an ensemble. These forces of postcolonial translation, which are at once embodied and technological, pervade the marketing of the mobile in contemporary popular culture. While the marketing of the mobile focuses on this contingent and intensive power of association, its design, interface, operating system, surveillance and security functions, and data-mining capacities seek to control and extract value from the embodied and interactive capacity to affect and be affected. But these attempts at control are always exceeded by new jugaads. This reduction of interactivity in the images and practices of mobile phone companies causes us to wonder if something like actual contingency is in fact possible. As Ash usefully notes, contingency is a "relative concept" as "the unpredictable and, thus, unknowable, taking place of an event. In this sense, contingency is not absolute, but relative to humans' understanding of a given entity or process. . . . Contingency is the unexpected happening of an event, whether that contingency is understood as absolute in the universe or simply relative to humans' incapacity to adequately understand it" (2010, 660).

Ash echoes the Spinozist contention that affects constitute the body's "power of action"—its unique capacity to affect (and be affected by) the world of bodies, forces, and things that it encounters. If affects are more than mere feelings or emotions, it is because they involve the constitution of immeasurable and unpredictable action-potentials, or an individual's dispositional orientation to the world as a modulation from state to state (see also Ash 2012, 2013, 2015b; Deleuze 1992a; Massumi 2011; Povinelli 2016). From the perspective of this method, every encounter subtly transforms an individual's affective capacities, either to enhance that individual's power of acting or to diminish it. Affect, therefore, describes both the distinctive set of feeling states actualized within a particular place as well as the store of action-potential of expressions, (in)capacities, and practices experienced in that space. The interplay of myriad practices and encounters, each with its own affective pitch and echo enfolded into these two experiences of affect—the feeling states generated in place as well as the durational capacities and ecological practices that each place makes possible—gives form to the diverse affective atmospheres discernible in place (Duff 2010, 885; Thrift 2004).

Affect, in this way, serves as a kind of map or tool of navigation whereby individuals negotiate lived space in search of those sites that later become places in and of the practices they support. Affect is the strange attractor lingering in place awaiting its realization in practice, habit, and sensation (Ash 2012, 2013, 2015b; Malabou 2005, 2008; Massumi 1993, 64–65, 2002; Ravaisson 2008). As Elizabeth Grosz parses it,

> Sensation is that which is transmitted from the force of an event to the nervous system of a living being and from the actions of this being back onto events. Sensation is the zone of indeterminacy between subject and object, the bloc that erupts from the encounter of the one with the other. Sensation impacts the body, not through the brain, not through representations, signs, images, or fantasies, but directly, on the body's own internal forces, on cells, organs, the nervous system. Sensation requires no mediation or translation. It is not representation, sign, symbol, but force, energy, rhythm, resonance. (2008, 72–73)

FIGURE 1.1. Sharman Joshi in the Airtel ad (2009).

In ecologies of sensation, one is affected by place before one might be affected in place through one's practices and habits (Ash 2014; Dewsbury 2012; Duff 2010, 892). What jugaad effects is a relative deterritorialization in practice, habit, and sensation, through which action-potentials become important in the formation of qualitatively new connectivities and the becoming of capacities. In its moment of capture by market forces, this new connectivity is what the Product Development and Management Association (PDMA) describes as a product innovation opportunity (see note 1 for this chapter).

Pursuing this line of analysis, let us consider an Airtel ad campaign concerning mobile internet access. The ads feature the popular Bollywood actor Sharman Joshi. The upper-caste, light-skinned Joshi carries with him a certain easygoing, yuppie-college, Hinglish-speaking metrosexuality, a Hindu-light masculinity that is associated with his generally convivial and humorous film characters (e.g., in *Style* [2001], *Shaadi No. 1* [2005], *Golmaal: Fun Unlimited* [2006], *Three Idiots* [2009], *Toh Baat Pakki!* [2010]). Each of these associations in different ways is important to the style of jugaad that these commercials offer to potential and brand-loyal consumers.

One ad begins with a familiar shot from mainstream cinema: a young student (Joshi) stands in supplicating attention before a distracted yet stern principal, who is dispensing patronage and signing various papers while seated behind his massive desk.[5]

JOSHI: Sir woh . . . festival ki dates? [Sir, what about . . . the dates for the college festival?]

Principal: Pardne aate ho ya festival manane? [You've come here to study or celebrate festivals?]

J: Pardne, sir . . . [To study, sir . . .] [*He looks down, embarrassed. As if by chance, his eyes are drawn to books on the table written by the principal on thermodynamics. He pulls out his Nokia phone and does a Google search for articles and books written by the principal. The connection speed is blisteringly fast, and he downloads the information instantly. The principal, busy with his paperwork, doesn't notice.*] Aapki thermodynamics pe saari kitaben pardi hain, sir. Aur aapka application of thermodynamics in zero gravity . . . what a point sir! [I've read all your books on thermodynamics, sir. And your application of thermodynamics in zero gravity . . . what a point sir!]

P: [*Impatient more than flattered*] Festival ki dates mail kar dena. [Mail me the tentative dates for the festival.]

J: [*Emailing from his phone.*] Bhej diya, sir. [I've sent them, sir.] [*A beeping sound comes from the principal's desktop, then we see a shot showing "New Mail." The final scene shot is of Joshi standing on the other side of the desk, leaning over the principal and pointing to the computer screen.*]

VOICE-OVER [*with the feeling of a kahavat, proverb*]: Baat karne se hi baat ban thi hain. [The deal [or harmony or agreement] is made only by speaking.]

In the final scenes of the ad, Joshi, the jugaad master of information, advises the "Old School" tyrant on his monopoly of discourse and authority. What constitutes the sense, value, and force of this symmetry-breaking event (DeLanda 2002; Deleuze 1994)? Its sense is charged with the intensity of inequality, of desire or recognition, of competition, and of capture; its value is measured in the intensive duration of the qualitative multiplicity of a hesitation into a diversion; its force is registered in the jump cut to the final shot: the student has moved over to the principal's side of the desk and is directing him to the appropriate window on his PC.

Harrison provides us with a way to understand this scene of hesitation and jugaad in terms of affect and duration. In this ad we encounter a form of interruption; the jugaad is potential and emergent. The sequence could be mapped as an anxious readiness, a communication breakdown, a gap opens up, and then an emergent order takes place. The mobile phone, the internet connection, the books on the desk, and the information online affect each other across thresholds of turbulence and protocols of translation (Ash 2013; Bhabha 1984). Harrison, drawing on Francisco Varela and colleagues (1991, 325–329), suggests freezing the sequence at the moment in which we have discernible hesitation, at the point where a turbulence of difference affects the action-potentials of human and nonhuman actants: "It is possible to say that we are at an interval: in-between stimulus and response. What is going to happen has yet to be determined and, further, the manner in which this moving on will occur 'is neither externally decided nor simply [internally] planned'" (Varela 1999, 329; cited in Harrison 2000, 503; see also Downey 2013; Latour 2005). The interval of the hesitation in the jugaad moment is strictly speaking an analogue in that through its duration a translation and synthesis of energy between different material and phenomenal states of being are actualized. These processes occur between sensory organs and perceptual systems within the body (proprioceptively) and between bodies and other objects/capacities (affectively) (Ash 2010, 661; Massumi 2002). For Harrison, it is habit that surrounds such encounters and resolutions, and "it is habit which is interrupted in intervals; from the embodiment of habit a consistency is given to the self which allows for the end of doubt: embodiment" as a configured and configuring feedback (Harrison 2000, 503; Ravaisson 2008).

The diagram of the relations of motion assembled in these commercials correlates habit and interval, not in terms of opposition or resistance but rather as intercalated processes of repetition and difference (Bohm 1980; Deleuze 1994). These ads tell us that the mobile is an intimate device that aids in rendering one's life radically flexible—at any time, you are free to connect to your passions passionately. The mobile is lived at the speed of life—no matter what life throws at you, the mobile enables you to seal the deal, be in harmony. The mobile jugaad image bodies forth a space-time of potentiality and contingency; in its material relations with its compositions, capacities, and milieus the image habituates a continual monitoring of the durations of encounters and the plasticity or modularity of spaces. These Airtel ads draw their surprising force from the image of the mobile as a device that is fundamentally stochastic, or unpredictable, because it is involved in processes that are open to chance, duration, temporal rhythms, interaction, improvisation, and variation of digital connectivity—given a certain infrastructure of bandwidth, signal coverage, information auditing, and battery power. It is a technology to jump over obstacles, or to jugaad. Joshi responds with "josh," as they say in Hindi—with some panache—and it is the mobile that enables him to effect his jugaad (an improvised trick). The mobile is represented as a kind of autonomic technology, as if inhabiting a liminal, threshold area of an extended machined neurology, where indeed anything, any act of creative innovation is possible, where plastic durations are open to becoming impermissible processes. The mobile, with the right branded connection, is pure potential, these ads tell us. It is impossible here to tease out ideology from form; the abductive intuition (Bergson 2012; Peirce 1995) of mobile jugaad is the patterned break—durational affect as a passage from one state to another—and technology of value capture cutting across the mobile phone ecology.

What we see in the Airtel ads is a specific set of images functioning as sensory-motor circuits that aim to prime the body for the stochastic dynamism of mobile technology, to gear up the body for that singular moment when *baat ban jayegi, hum jugaad kar lenge*, the deal will be sealed, we'll play a trick. The jugaad expresses a certain unparalleled joy—of getting away with something through sheer ingenuity, possibly an extralegal wager? The joyous force of jugaad, its durational passage, that Airtel is trying to appropriate to its brand equity can be seen as an attempt to brand a cool "stochastic connectivity"—the force of dynamically responding to life's chances, as if mobile connectivity were a time slip into the future.

On the one hand, I would affirm that contemporary analyses of affect and digital practices in mobile ecologies need to take seriously the capacities of the body and the capacities of contemporary mobile technologies as a coevolution. As such, a postphenomenology (Ash 2013, 2014; Massumi 2002, 2011; McCormack 2005; Stiegler 1998) of mobile-human connectivity would aim at finding regions of resonance and dehabituation between them, diagramming their intensive variations to maximize the emergent capacities of this assemblage for a "nonbranded joy." A nonbranded joy would be one that is common and noncommodifiable. As such, the mobile phone-human perception assemblage could in fact become a time slip into the future. The branding, the spatialization and capture of the mobile ecology habituates consumerist clichés of connectivity, invested as they are in implanting sensory-motor circuits that yield the greatest "average revenue per user." Thus, Airtel's image of digital cool projects in a partial way what happens to our machinic, networked bodies within variable states of things that tap into our bodily capacity to sense and be affected.

Conclusion

This chapter explored the limits and possibilities in the method of understanding the affect of jugaad as ecology and event at the intersection of lived space (the mobile phone ecology) and emerging markets (macro and micro) in India. I have tried to show, first, that jugaad as subaltern social practice and as business model are irreducible to each other. There will always be something lost in translation. The concept of postcolonial translation mutates how we understand relations of power between bodies and ecologies; it allows for a specific thinking through of the political stakes of the capture of subaltern tactics and strategies by monopoly interests. Postcolonial translation becomes in the process the production of ontogenetic common notions: jugaad as commoning tactics for undermining private property's hold on everyday life (Deleuze 1992a; Hardt and Negri 2009). This analysis shows that while there is a certain complicity between jugaad as everyday tactic and the ad hoc infrastructures of developing postcolonial economies, the translations and syntheses necessary to couple functionally technical objects to human perception and nonhuman ecologies expands practices of human resistance, displacing its binaries, by diagramming action-potentials distributed across assemblages (Povinelli 2016). As both technosocial practice and corporate strategy, the jugaad passes through the machinic phylum, increasingly as a pedagogy of risk and logistics in algorithmic capitalism, but more often simply as an experimentation with any instrument that is most conveniently to hand, for instance, a twelve-volt battery (Rossiter et al. 2012).

Jugaad practice complicates much writing that "argues technology inherently or absolutely closes down the potential of the future (as inherently contingent) by framing and calculating the future as a temporal mode of the present" (Ash 2010). Mobile phone and mobile phone provider companies in India present the environments in which users operate devices as both controlled (constant access to electricity, bandwidth, and signal) and contingent (the device allows a given jugaad to move consumer experience from stasis to improvisation). The aim of this controlled contingency is to increase the potential for "positively affective events to occur in the future"; however, given the complex ecologies of information, energy, matter, desire, and technology that constitute mobile phone cultures in India, none of these events can be predicted or totally captured in advance (Ash 2010, 667).

Finally, this method allows for effectively analyzing the affect of jugaad in its phenomenality, that is, as a process of ecological emergence. As Povinelli assembles it in *Geontologies* and as recent work in feminist organization studies affirms, affects always travel through interrelated assemblages composed of some form of matter or another (Povinelli 2016, 149–150; Pullen et al. 2017; Fotaki et al. 2017; Linstead and Thanem 2007). Echoing our analysis of postcolonial affect but from a posthuman context, Ash suggests that through these coupled atmospheres "affects are a matter of force as much as any kind of content. Thinking through this associated milieu also allows us to consider how the same affective force has differential impacts dependent on the body or entity it encounters and how single objects can create fields and atmospheres of affects that, in turn, generate spaces" (2015a, 5). The object or event of jugaad becomes unthinkable when abstracted from its ecology; indeed, both object and event become parts of common notions relating ecology to ecology. For Ash and others in this field of affective ethnography, an event is an effect of a material assemblage of various entities, forces, and rules "working together to encourage and prohibit specific forms of movement and action" (Ash 2010, 668; Povinelli 2016). Situating the jugaad event in an ecological frame is useful in understanding how the potential for events to happen is designed into digital and physical environments. More, the idea of ecological emergence allows for an understanding of how various bodily states (such as anticipation and boredom or pleasure and pain) "can potentially be produced and controlled through manipulating affective relations in the environment. This then allows us to interrogate the possible responsibilities the designers of such environments have in the kinds of affective relations (and thus bodies) they (potentially) construct" (Ash 2010, 668). The affective relations mobilized and potentialized in jugaad are instruments and targets in mobile phone business strategies, while subaltern practices of jugaad continue to disrupt value and exceed this capture. Thus, returning to the political stakes of such an analysis, there is an important difference not just in material conditions but in value itself between the jugaad practices of ordinary Indians, which beyond just a way of fixing things is quite directly tied to the persistent effects of (post)colonial struggles against hierarchy, authority, and deployed power. The uptake of jugaad in contemporary mobile phone business practice as well as the makeshift ecologies through which particular jugaads emerge show the ongoing force of these struggles. In this view, jugaad inherits a nonhuman capacity for subaltern resistance and this (postcolonial) resistance remains latent in the space it makes open to improvisation, even if this space is daily captured through technologies of digital control, more is made and capture is never total or assured.[6] As we have seen, at some moments the value that corporate mobile phone companies want from jugaad and what ordinary Indians want in a jugaad can line up, as in wanting internet access cheap and fast, but this is a postcolonial translation of the contingency and ecology of jugaad, a social synthesis specific to postcapitalist India, but related to others throughout the world. In closing, I note that many of these mobile carrier ads are no longer available on YouTube, but what is readily accessible are tutorials on how to tether a phone to an Airtel data network for free.

NEOLIBERAL ASSEMBLAGES
OF PERCEPTION AND
DIGITAL MEDIA IN INDIA

In this chapter, I diagram in more specific detail the neoliberal forms of work and workaround practices in India. Through close attention to the habituations and value capture in what is commonly referred to as "business services/processes outsourcing," I develop an analysis of the historical and material conditions of what today is perhaps better known as affective, immaterial, or communicative labor. The perceptual schema of the value-adding digital image is often, but not exclusively, associated in many parts of India today with various forms of jugaad. I develop here my counteractualizing diagrammatic method, which brings together different forms of communicative labor, digital image technologies, and the changing capacities of the body, or affect. As we saw in chapter 1's analyses of media practices, affect, because of its relational, emergent, and potential character, is a crucial site of modulation, control, and new strategies of economic value. Jugaad practice expresses this struggle, as do contemporary feedbacked media ecologies. Numerous feminist investigations, analyzing the potentials within what has been often designated as women's work, have grasped affective labor with terms such as kin work and caring labor. These analyses open a new set of questions in considering the affective passages of jugaad: what perceptual shifts emerge in and through the jugaad event, and, following Elizabeth Povinelli (2016), what forms of attention begin to matter in jugaad ecologies? Interweaving interviews conducted with jugaad practitioners in two of India's supposedly shining "smart/jugaad/Big Data cities," Bangalore and Mumbai, with analyses of the documentary *Office Tigers* (Mermin 2006), I further consider the political and economic contexts of the emergence of a new dominant sensorium, or ecology of sensation (Adorno 2013; Benjamin 1999, Buck-Morss 1991; Clough and Halley 2007; Puar 2017). Indeed, the emergence of a neoliberal digital perception in the South Asian context through pervasive and continuous processes has coevolved with the overall "informatization" or datafication of various forms of everyday life and formal and informal work. Of course, India after liberalization does not present us with a clear-cut transition from an analogue postcolonial disciplinary to a globalized digital control society. Regionally diverse and complexly embedded, Indian forms of power and valorization are mixed and have evolved nonlinearly in postcolonial hybridity. The mutations of jamming, piracy, and viruses are effects of the processes involved in the reproduction of hacking ecologies. In South Asia, the indigenous practice of anticolonial insurgency through subaltern networks of rumor and its associated dangers of unpredictable native movement provide a complex historical terrain with which to understand the movement of a virus, plague, or hack (Guha 1983; Spivak 2012). This chapter considers the capacities of these new mutations in the affective environments of the digital image.

Hacking Habits and Potentials

Affective ethnographies of contemporary digital media in Mumbai, Delhi, and Bhopal have become more probable with the very rapid adoption of the mobile phone, tied as it was to increasing social and economic pressures after globalization to increase national, household, and individual productivity and to secure cutting-edge connectivity networking various information platforms (internet, governmental, regional, gaming, etc.; see Hardt 1999; Hardt and Negri 2009), which displaced voice telephony as a default mode of communication. In India these mobile and digital connectivities ingress into more and more intercalated bodily and cultural processes.[1] Given these dynamics, the human-mobile assemblage in fact is a historic potentialization of actualized forms of habit in Indian cities (see Ansell-Pearson 2001; Bergson 1988; Deleuze 1988a, 1992a; Grosz 2013; Heidegger 1977; Malabou 2008; Nietzsche 1999; Ravaisson 2008). In other words, perceptual ingression opens the human onto nonhuman worlds of potentialized becoming as the increasing quantity (intensive gradient) of connectivities contracts into a qualitative change in perception, memory, and sensation; it is precisely this potentializing of the habituated body that has been the target of contemporary neoliberal entrepreneurial value strategies in the mobile phone industry in India. This dialectic between freedom and control at the level of discourse functions to obscure a more fundamental set of intensive and informational processes that are transforming the embodied experience of mobile networks.

These experiences are shaped by gender, class, communal, and sexual identities, social and economic norms and infrastructures, structures of feeling, and aesthetic styles (see Nandy 1998; Spivak 1999; Sundaram 2009; Williams 1977). Together, questions of identity are bound up with the habits and styles of the body, and the historically and materially specific sensoria, or ecology of sensation that one affirms and/or negotiates day to day is the virtual and actual "fold" of a body's coevolving difference. Jugaad practices, as ways of working around different forms of power, affirm the becomings that potentialize this difference.

Before we proceed further, I want to consider some of the contours of the present debates in the overlapping scholarly fields of feminist ecosophy, media studies, and postcolonial autonomia (this last is an emergent conjuncture). This chapter engages with the forms of analysis of media ecologies that draw on both interpretive and empirical research into the changes in perception and consumption in digital cultures in India. The work on critical media ecologies (Balsamo 1996; Brosius and Butcher 1999; Clough and Halley 2007; Fuller 2004; Hansen 2004; Haraway 1991; Hayles 2005; Manovich 2001; Massumi 1993; McLuhan 1994; Menon and Nigam 2007; Parikka 2012; Parisi 2013; Puar 2009, 2017; Sundaram 2009) has provided specific tools for this analysis. The first is the postcolonial and queer feminist analysis of the body-in-media. The cyberfeminist tradition that continued to develop the research into the cyborg begun by Donna Haraway and Anne Balsamo, the work on affect as capacity in the feminist materialism of Elizabeth Grosz (2013) and Patricia Clough (2018), and the queer assemblages of Jasbir Puar (2017), Geeta Patel (2017), and Elizabeth Povinelli (2016) have been involved in the articulation of a new feminist politics of the technobody. This new diagrammatic method does not limit itself to analyzing the female body in the patriarchal regime of code (Kristeva 1982), or to the modes of control tied to biopolitical distinctions between life and nonlife (Povinelli 2016), but rather involves these critiques in various modes of counteractualizing control and subjugation, and further experiments in the creation or freedom of heterogeneous bodies-in-becoming. Anna Munster powerfully draws on this tradition as she writes, "At one limit or pole, we find the potential directions in which a flow of matter moves or can be organized. Here, a set of exchanges is laid out that describes all of the possible relations a flow might enter into: what capacities and functions allow the movement of this flow into a specific or more localized material formation, such as a particular technical machine" (2006, 13).

A second important tool emerging from diverse literatures is a focus on habit and its ecologies. In the work of critical theorists of modern mass media, the repetitions built into the routinization of work after Fordism, and the generalization of information technologies across all forms of labor and consumption, have collapsed the world of work and leisure, and consequently brought more and more once "private" habits into the realm of capitalist value, organization, and forms of personhood (Berardi 2008; Fleming and Sturdy 2009; Hardt and Negri 2009; Peticca-Harris et al. 2015; Pullen and Rhodes 2015; Safri 2015). In the example I discuss here, an iPhone photography habit, a certain communing with nature (a kind of animism common in different spiritual traditions in South Asia), and the confrontation with the limits and possibilities of ego-in-money, are considered from the perspective of its ecological flows and patterns of attention. The increasing investment in such processes signals a global shift toward affective labor and toward "caregiving" as a decisive form of value production in post-Fordist capitalism and postcolonial liberalism. Drawing on Michel Foucault's work on "human capital" (2007, 2008) this critical tradition sees the formation of habits in the realm of the digital closely tied to new emergent forms of home-based labor, entrepreneurship, and social networking (i.e., the joyous passions of self-exploitation).

Finally, the contemporary feminist theorization of affect, for instance in work by Patricia Clough, Jasbir Puar, Elizabeth Grosz, Sara Ahmed, Lauren Berlant, Jack Halberstam, Stefano Harney and Fred Moten, Elizabeth Povinelli, Angela McRobbie, and Luciana Parisi, has greatly benefited our understanding of information and communication technologies as assemblages of perception, sensation, habit, information, capital, and technology. In this framework, women and men are understood to develop capacities to affect and be affected by feedback loops with geological, technological, and social ecologies. This has allowed for a nonlinear dynamical approach to media habituations that would consider the ecological effects of contingency, chance, and transformation as well as a critically realist analysis of forms of power that operate through the ecologies themselves.

What are the implications of using nonlinear dynamics for media studies? There is, of course, the danger of cultural critics dabbling in science, a danger that Deleuze warned of in his prefatory remarks to *Difference and Repetition*. Science, in this case dynamical systems theory, is not used here to legitimate the normative claims of a critique grounded in truth and the good; rather, it is used as a methodological resource in practices of exiting from the androcentric worldview of much of contemporary discourse on media (different sociologies of media).[2] For now, let us note merely that the analysis of perception, affect, and habit in mobile media is tied to the dynamic of interrelations among complex distributed informational and bodily processes. This takes ecology out of the human body, or out of its skin, and indeed sees the skin itself as always already folded into, distributed across several multiplicities at once, and so it affects in turn their ecologies.

In terms of method, media assemblages trace or diagram ingressions of matter, energy, and information into intensive ecologies of image, sound, tactility, movement, and sensation. This requires situating together, through strategic parataxis, the coevolving or feedbacked processes of embodied sensation and information flows from which various forms of habit, digital media, institutions, and industries emerge. Digital networks establish dynamic interfaces that positively feedback through sets of recurrent media events, and from these processes assemblages or correlated compositions of sensations, matter, technology, neuronal processes, brands, and populations take on definite forms. As quickly captured objects in entrepreneurial strategies of control and monopoly rents, novel value streams emerge by modulating the affective capacities of the body (the variable ability of the body to affect and be affected), and quickly they enter into new algorithmic ecologies and networks of circulation (Foucault 2007).

Contemporary forms of communicative labor, such as digital imaging or the call center, have been termed "affective" or "immaterial" labor by various commentators. In India, the employment of affective labor has been fundamental to the rapid transformations in the country's globalized economy. By focusing on bodily capacity coevolving with digital technology, the diagrammatic method of ecosophy shifts critical attention to gradients of sense, value, and force in historically specific assemblages of technologies, bodies, and populations. In India these coevolving bodily capacities are undergoing critical thresholds of change, given the generally high level of uptake of digital mobile technologies and information and communication devices more generally throughout the country.

Notes from the Smart City's Undercommons I

Consider as another approximation in diagramming these overlapping ecologies this edited excerpt from an interview conducted with an upper caste, upper middle class, male business processes/services quality-control worker in Mumbai.[3]

Pritish is twenty-eight; he was born and brought up in Vashi, Navi Mumbai. He lives with his mother and elder sister. His father is deceased. They live in an extended two-story CIDCO[4] accommodation, which comes under Mumbai's municipal housing now. (For an explanation of the CIDCO designation, see note 4 of this chapter.)

Pritish has worked as a quality analyst for the past three years for Tata Consultancy Services, in a "special economic zone" in Powai, an Indian-based, multinational information technology (IT) service, consulting, and business solutions company, headquartered in Mumbai, Maharashtra. He has been working in general since the age of eighteen. He attended high school in Vashi. While attending Vivek College of Science and Commerce, Chembur, he decided to quit formal education and start full-time work. He has eight years' work experience of working in various business-process outsourcing (BPO) companies in Mumbai. He doesn't believe Mumbai is a smart city (a "fail-safe city") for all its citizens; within the same city he sees some people using plastic economy and smart technologies, while others struggle in daily life. Pritish is an avid traveler, and he frequents the mountains of northern India. He is also fond of phone photography, recently having purchased an iPhone 6. For him, there is no limitation on what you can do in the city. "This city is on everyday, it does not have a switch off button, it is always running." In his interview with Rachna Kumar, he elaborates:

I am a twenty-eight-year-old man. Been in this city all this while, exploring it. Still trying to find the meaning of life. I work at a multinational company, as a quality analyst. I was raised by my mother, she's been a single parent since I was age three. My mother is a housewife now; she's quit working. She is a cancer survivor. My sister works in another multinational company, an airlines. I work in a department of an Australian client's telecom department, where I have people processing orders for telephone and internet for the provider in Australia. My job as an analyst for quality is to audit these orders and make sure that there are minimum errors in the orders processed so our customers are not hampered. So on a daily basis I would monitor calls that they take from customers and orders that they process on the system. Yes, lots of communication work, and that's the best part. I love meeting new people. My job involves speaking to at least forty to fifty people daily; usually it would be the same people all year round, however, there is ample opportunity to meet new people as we communicate with other teams as well. The job keeps me on my toes, and also I fulfill my wish of meeting new people everyday and trying to break the monotony in the job. I make close to six lakhs; it fluctuates as we receive a nightshift allowance. I live comfortably. I have a good paying job. I have all the necessities and luxuries needed for a man of my age living in Bombay.

Mumbai has always been thought of as a land of dreams; the opportunity to make a quick buck is high. People come from all over; this city is very providing. In Bombay if you live there, you have to be a contributor to the city. If you eradicated the slums from Bombay, the city would come crashing down. It's both the provision made by the city, mainly the politicians, who want to name an area after them; in the slums they can order online and get delivery to wherever they are. Advertising causes need in society, right? Today the market is taken by storm by the digital, this online provision. If you did a market study, you would find a large number of people checking online with their mobile to explore every realm of the digital world we have created; it's not a small sector of society doing it, it's *everyone*. Everyone has a mobile phone today and everyone has the means to explore the entire realm of this digital world that we have created. It's no longer for the elite. I remember in the '90s or early 2000s when the mobile phone industry was really booming, but it was always a thing for the elite, if you had the money or you had some position in society then would you then have a mobile phone or a computer. Me being born in the late '80s, I was lucky to see the transition of this world from relatively no technologies to an age that is highly dependent on technology. Even Bombay I see transitioning from a no technology society, manual labor society, to a highly digitally literate society. Even homeless slum people who were computer illiterate are getting training and integrating digital technologies into their businesses. Take a cable TV operator in Dharavi, right? Initially it was simple being a cable TV provider: you had a cable into the home and it was pretty straightforward, now with all the set-top boxes coming in, Tata TV types, and the government backing it, making everyone get a set-top box, though 99 percent of the people don't know about what the technology actually is.

[Q: Do you have a smartphone?]

The smartest dumb phone I have, it's the iPhone 6; not following normal norms [sic] of getting an iPhone and getting that position or, um, what do you say? That place in society that says now you have an iPhone that means you are doing well, you are rich, you are a man of current times—none of that, it was only for the camera. I love the camera on the phone. Today I see people buying iPhones to show others they have one, that they have status; they wouldn't use 10 percent of its functionality; they don't know shit about the phone, they get a loud ringtone and flash the logo every time they talk on the phone; so everyone sees and then they build up a perception around it. But I'm not that kind of guy!

I do a lot of nature photography. I don't take selfies. That's for people who have accomplished nothing, but are documenting everything about their stupidity. We are a generation who has accomplished nothing at all, but are documenting it like crazy. We have terabytes of our own pictures, getting drunk and doing shit with our friends. I would love to record my experiences in the material world, so I can relive those inside my head. So I don't photograph people. When you are traveling, people don't fulfill your experience; nature fulfills your experience. We are empty inside, we are living in the city chasing this dream, and it's a race already fulfilled. Running after external things, trying to get things on the outside, and we are empty on the inside. I find more important things discovering what is inside myself when I am in nature. Living in the city—in this thriving life—all that seems good, but it's still an empty life. It's when I travel outside the city I realize that all I have to offer nature is my money; nature doesn't take money; nature doesn't love my money or my clothes or my iPhone. Nature doesn't love me, the person I think I am, the person my ego built. Nature doesn't want it, it just wants my energy there, if I can be a contributor. That's why even in the city I look at people as whether they are contributors or they are leeches. The reason why there is such a good union of different sections of society in Bombay [is] because everyone is a contributor. Even the downtrodden here are ok with it as long as they are a contributor, and the rich live with the poor all together. My city is only as strong as the weakest section of society. Here the weakest section of society is so smart—they won't stay homeless. Tallest buildings and deepest slums. Here in Bombay you won't find apathy in this city. You see Africa, you find the same conditions as the slums here. Those people and these people are completely different. These people will still go to the mall and have a good time on a weekend. And then get back to the life of saving all month until they have another chunk of money again to spend on the luxuries that the city has to offer. Downtrodden people are not giving up. They are thriving, they contribute to this city, and that's the beauty of this city. That makes it a smart city.

The form of labor that Pritish engages in, broadly understood under the heading of cognitive or immaterial labor (Lazzarato 2006; Hardt and Negri 2009), has certain key characteristics tying it to neoliberal regimes of accumulation. Consider how the indirect pressures that elicit and sustain long work hours among technical, professional, and managerial employees, ostensibly through personal choice, can also be understood as a neo-normative, biopolitical control mechanism (Adams 2014; Ajana 2013; Amoore 2013; Arboleda 2015; Fleming and Sturdy 2009; Peticca-Harris et al. 2015, 573; Pratt et al. 2017). The norm that had encouraged workers to conform to a rigid organizational rhythm and distribution topologically contorts into a sieve with a variable mesh as workers are now extolled to "be themselves" and bring more of their "authentic selves" to the workplace (Deleuze 1992b; Fleming and Sturdy 2009, 570). As several have noted, this exhortation for "creative" workers to be individuals and authentic selves at work goes hand in hand with the expectations for workers to be individual entrepreneurs in the market, to be self-reliant and shoulder the risk of maintaining their own employability (Peticca-Harris et al. 2015; Webb 2004; Fleming and Sturdy 2009; Berardi 2009b). Within this context, having "real" fun at work and building authentic, positive social networks while individually interpreting and enacting the professional norms of one's occupation are key signals of careerism and drivers of career progression (Peticca-Harris et al. 2015, 573).

In Pritish's worldview, each of these labor dynamics operates both its fetish (sensory-motor circuit as image) and its habituation (using a phone's expanding functionalities). For instance, we must consider the connection between the importance he places on his home as emotional resonance and property value (at one point he notes that his home is worth around 1 crore rupees) and his disavowal of showy consumerism (I'm not that man!) and the care he wants to give to "nature" expressing itself in his loving photography of landscapes; or the connection between his acute consciousness of the image machine of contemporary Western media (the iPhone newbie) and his reductive understanding of misogyny in the urban (at one point, he says, "Before there was no crime against women in Bombay. This increasing incidence of rape is because of the media's portrayal of incidents of rape"). In Pritish's work-self-expression ecology, an abstract diagram functions before and beyond the identification of resonances as similarities. Such diagrams diverge from themselves in the actual and virtual circuits of these ecologies, through its phase transitions, and from the emergent affects they both release and enfold.

Contextualizing the Emergent Digital Ecology of Sensation

I turn to consider the ecological effects of the breakup of various state media monopolies such as Doordarshan TV and All India Radio, and the emergence of India as a global power within neoliberal "casino" capitalism (Boddy et al. 2015). Institutional reorganizations and market liberalization emerge with and through changes in consumers' sensorium, itself an increasingly datafied and complex web of feedback loops to global logistics, circulation, production, value, and labor. In other words, the correlated rhythms of the body—rates of perception, forms of attention, intensities of sensation, autonomic processes, and proprioceptive capacities: in a word, habit—shift as the speed, scale, and patterns of media themselves change.

The past twenty years of Indian media history have seen the catalyzing, reinventing, capturing, corporatizing, "revitalizing," and mutating of already vibrant print, visual, craft, and oral media practices that long predated the coming of electronic (colonial) and digital (neoliberal) media.

These practices continue in a relatively independent and still-evolving sphere, one that has many points of connectivity and feedback into the new media. What are the affordances of actually existing South Asian public spheres as mobile digitization becomes part of the neural plasticity of the urban? And in what way is this both a political and ethical question?

Several contemporary neurophilosophers have developed critical work on the extended brain under conditions of ubiquitous computing, which can help us situate what is at stake in developing these two correlated aspects of mobile media assemblages. This conversation around the extended brain helps us to situate the dynamic geographies analyzed by Doreen Massey (2004, 2005), Bruno Latour (2005), Paolo Virno (2003), Andy Clark (2003), Elizabeth Povinelli (2016), and Jane Bennett (2010) in the newly networked plasticity of India's smart and anti-Dalit cities. It is this multiphylum plasticity, whose encounters, assemblies, détournements, contractions, and accelerations mark the potential and actual affordances of mobile-human assemblages. We will return to the question of urban and neural plasticity in chapter 3.

Specifically, extending Clark, we can diagram our home and office environments becoming progressively more intelligent as well as copyrighted, "courtesy of multiple modestly powerful but amazingly prolific intercommunicating electronic devices," so that capitalist computation becomes embodied and distributed throughout space. In this techno-ecology of habituated sensation, interfaces multiply, become naturalized, and are "rapidly invisible to the user." Once this happens (after a critical phase transition), the activities and interactions of this ecology operate as the unremarked backdrop on which the biological brain and organism learn to depend for the infrastructure of its habituations. Human cyborgs in the age of mobile, ubiquitous computing (and its attendant habits) make the most in terms of monopoly rents of cortical plasticity through cognitive extension. The implications of this coevolution of capital and neurology are profound: "Various kinds of deep human-machine symbiosis really do expand and alter the shape of the psychological processes that make us who we are. The old technologies of pen and paper have deeply impacted the shape and form of biological reason in mature, literate brains. . . . The moral . . . is simply that this process of fitting, tailoring, and factoring in leads to the creation of extended computational and mental organizations: reasoning and thinking systems distributed across brain, body, and world" (Clark, 2003, 32).

Today, in much of the global North and in Indian rural-urban networks of caste-based communication, smartphones are introducing users to new waves of partly invisible, partly hypervisible, user-sensitive, semi-intelligent, knowledge-based electronics, feedbacks, interfaces, and algorithms; indeed, some have claimed that in North America and Western Europe these phones are poised to merge "seamlessly with individual biological brains," and in so doing they will ultimately "blur the boundary between the user and her knowledge-rich, responsive, unconsciously operating electronic environments" (Clark 2003, 34). In India, the 1980's subscriber trunk dialing revolution that changed the very conditions of communication throughout India was part of the broader set of innovations developed by the team of researchers brought together by the IT entrepreneur and government consultant Satyanarayan Gangaram "Sam" Pitroda. The subscriber trunk dialing cultures were the precursor to the mobile phone in its tendency to correlate movement and communication across and eventually beyond the nation. For another example, consider the nearly 150 major newspapers in 100 languages published in India daily and the continued importance of local and regional "mela" performances interfacing with different aspects of new media consumption under neoliberal India. In emergent mobile phone cultures, phone calls and short message service increasingly interrupt the patterns of social interaction, and these repurposed, viral audiovisual media flows are partly driving in turn the emergence of new media assemblages in radio, print, video, internet, performance, mobile phones, and art. It is in this postcolonial singularity that we must situate the rise of postliberalization digital media institutions and practices, their forms of organization, and their concomitant intensive, embodied infrastructures, or ecologies of sensation.

Jugaad Time comes out of the researches and methodologies diagrammed above. In what follows, I have three correlated arguments that I will advance through one main media example. First, there are specific feedbacked processes that must be analyzed in their emergence together in formal and informal mobile phone ecologies. This suggests the correlation of material flows of information, technology, value, and sensation. All these flows have singular durations and timescales, specific emergent capacities, and actual and virtual multiplicities: immense and immeasurable, but susceptible to control (Bergson 1988; Deleuze 1994, 1998; Hardt 1999; Negri 1999). This involves thinking the bodily capacity to both affect and be affected beyond the organic human body and beyond habituated emotion. Emergent capacities, for instance, are both potential and actual, given the tendencies and parameters of change in more or less dynamic networks of IT and human perception. This method involves thinking in terms of the body's variable (mis)connections with information in distributed networks. Of course, another crucial process is that of "subject formation" itself, through which new habits of consumption, pleasure, and perception form in relation to this new information ecology. It is useful to recall what this critical method has long said of the question of habit: habituations repeat and create in the same practice, in that mutations arise from the statistical regularities of the repetitions themselves. This is the new collective subject or multitude of Indian digital cultures: habituated and potentialized at once, a subject of continuous modulation and expressions of that modulated freedom. This subject has for some time now been the virtuoso of the jugaad. As we shall see, this flair for working around, or sidestepping, the regime of protocols (intellectual property and internet protocols) has now fully enveloped digital culture in India.

Second, the analysis of dominant media flows should foreground the production of values. Negri (1999) suggests that the relation between the globalization of value-added services, processes, and products (through, for instance, call centers or business process outsourcing), and the embodiment of communicative labor (in new habits of communicating, attending, and consuming) is a useful point of departure in such an analysis. Time-space compression in work patterns in digital networks and communicative labor, where the instantaneous delivery of information and the time lag between the business day in New York and Chennai intensify the nature of work through forms of always-on, extreme connectivity (Gascoigne et al. 2015; Granter et al. 2015). The embodied experience of work in BPO and the modes of interactive attention involved in consuming digital images generated by computer algorithms are correlated in the emergence of new information systems and their attendant ecologies of sensation. The stylization of contemporary Indian urban life as the canny practices of working-class/lumpen jugaadus in commercial cinema has offered up a new cinematic cliché: the "lonely bubble" of the distracted and privatized mobile phone user in recent Hindi-Urdu films, in which the value-added of the interactive mobile phone screen divides the cinematic scene, interrupting the narrative and enabling a forking away from profilmic time-space, a folding of time-spaces, their intensification along the lines of an indi-

viduated reality centered on consumption, security, and the entrepreneur-
ing of various populations: "reality" intensified. This suggests something
fundamental in the organization of media in India: as media has globalized, so habituated perception has become privatized, isolated, and newly
hierarchized. But this is in no way a completely one-sided process; there
are many mutinies against these dominant processes, and part of my aim
in this book is to demarcate the scope and scale of resistances to and cre-
ations away from commodified or monopolized forms of value capture and
its attendant control habituations.

Thus, following through on the insights of a queer cyberfeminism, the
method of media assemblage analysis considers the body's capacities with
flows of information and forms of power. Such an analysis draws on and
departs from much of contemporary scholarship on consumption in South
Asia. The argument outlined above suggests that, in India's emergent media assemblage, sensory-motor circuits functionally correlate with new
forms of networked and revalued flows of information that traverse media
platforms. This is especially clear in the emergence of what commentators
have called "affective labour." Keep in mind that for Michael Hardt and
Antonio Negri, affective labor produces "social networks, forms of community, biopower [where] the instrumental action of economic production has been united with the communicative action of human relations"
(qtd. in Dowling 2007, 119; Hardt and Negri 2009). My argument extends
this line of thought to suggest that business-process outsourced labor, as
communicative, relational, "creative," and hyperdisciplined, modulates the
rates and rhythms of attention and multisensory perception in the neoliberal workplace in India today. The aim of a media assemblage analysis
would be to diagram pragmatically ways of productively intervening in
such circuits of habituation by refunctioning and collectively communing
the material connectivities themselves (Linck 2008).

Throughout this chapter I draw on a set of analyses that has developed
the notion of affective labor as a decisive break in the organization of value
under neoliberal, globalized capital. Something radical is happening to
the experience of work in India; part of this is due to neoliberal regimes of
accumulation and their attendant intensifications, part to global shifts in
patterns of interaction through digital media, part also to ongoing forms
of autonomous exit toward noncapitalist commons. Through these mutually ramifying and antagonistic dynamics, affect-as-capacity enlivens the

question, Which work? (Weeks 2011). Feminist political economists and postcolonial critics engage the affects of archives, technology, worship, interaction, and communication; distinct from "emotion," affect is defined by its relational character, not limited by an internalized feeling. In that regard, affect is considered "preindividual," operating in those strata of being where the subject and populations meet (Massumi 2002; Shaviro 2014). Understood in this way, affects seem to be at stake everywhere within a labor world dominated by dynamic, distributed, digital connectivities, and by the imperative of building connections, of defining one's own personality—for instance, through iPhone nature photography—as a knot of multiple networks. In order to be successful in a world where labor is becoming increasingly flexible, casual, and "precarious," one has to show that he or she is capable of building relations, of producing affects. In a situation in which the boundary between friendship and business is itself being blurred (are you building a connection with a certain person because you like him or her, or because that person can be useful for you?), specific problems and disturbances arise (Boltanski and Chiapello 2007; Mezzadra 2008). As Michael Hardt has usefully noted, the productive circuit of "affect and value" has affirmed an autonomous circuit for the constitution of subjectivity in experimentations and practices alternative to the processes of capitalist valorization. These theoretical frameworks have sporadically but not as yet effectively brought together the postcolonial, Black radical, and subaltern studies insights of Gayatri Spivak, Homi Bhabha, Cedric Robinson, Dipesh Chakravarty, Vijay Prashad, Fred Moten, Ranajit Guha, Partha Chatterjee, and Edward Said with Marx and Freud and their Deleuzo-Guattarian interlocutors. In this ongoing project of reconstructing a transnational project of solidarities across forms of affective and manual labor, a reconceiving of the geopolitics of affective labor is afoot. Developing the strategic and practical resources for postcolonial, queer, Black and Dalit, and feminist politics in concepts, such as the (new) international division of creative or cultural labor, this work develops new methods for diagramming neoliberal desiring production (Deleuze and Guattari 1987). Significantly, numerous feminist investigations have analyzed the potentials within what has been designated traditionally as women's work and have grasped affective labor with terms such as kin work and caring labor (or "labor in the bodily mode") (see chapter 3). These analyses reveal the "processes whereby our laboring practices produce collective subjectivities, produce sociality, and ultimately produce society itself" (Hardt 1999, 89–90; see also Lovink and Rossiter 2007; Harney and Moten 2013; Saldanha 2007; Terranova 2004; Lury, Parisi, and Terranova 2012; Staples 2006).[5] How, then, in the context of the Indian anti-Dalit smart city economy, is the function of affective labor correlated in both business outsourcing and digital perception? More, how does this correlation play itself out in varieties of digital jugaads today? One modality of this evolving functionality is the nonlinear yet regulatory system of distributed digital networks; another is the modulation of subjectivity in the capacities of attention and sensation of value creation.

Media Assemblages and Sensation

Let us follow these molar tendencies and structures through a nonrepresentational, or diagrammatic, encounter with a 2006 documentary by Liz Mermin, *Office Tigers*. Now, by nonrepresentational I do not mean that representation does not exist or that it has been overcome, but that representation involve actualized forms of affectivity (sensory perceptions) abstracted, or subtracted, from the flow of sensation or perception itself (Bergson 1988; Nietzsche 1983). Following the changes in relational capacities (affect) in the dynamics and trajectories of this subtraction of digital perception from intensive informational flows is the first approximation of a diagrammatics of affect. In this, we follow what remains radical in American pragmatism, insofar as thought is not geared toward the stabilization of habituated grids of representation, but rather toward the becomings (deterritorialization and reterritorialization) of its ecologies (Massumi 2002, 2015a, 2015b; Povinelli 2016). Thus, representation understood as affection-images is involved in the dynamic affective charges and changes of the body's affordances (capacity to affect and be affected) in specific but connected regimes of capitalist valorization. Second, by diagram I mean something like a "plan" in the sense of an intersection of vectors, or processes of composition, which suggests that diagrammatic thought is ontologically implicated in ethical experimentations in informal digital ecologies (see Deleuze 1986; Halpern 2015; Parisi 2013; Wark 2012). Situating the sensory-motor circuits of neoliberal capitalism's emergence in the early 2000s in India, my analyses here suggest that India, poised at several phase transitions at once, is in the process of a radical transformation of the time-spaces of the smart/Big Data city.

Office Tigers is a rare, relatively early glimpse into the life of a business service outsourcing enterprise in India. Its historical specificity shows us that the advance of neoliberal and neo-Hindutva agendas linking security to purity through technology has adapted the digital habituations that had developed in the first decade of the transition (1991–2001). The documentary revolves around Joe Siegelman, a thirty-four-year-old American ex–Goldman Sachs executive, who authoritatively directs the movement of the camera around the Chennai offices of Office Tiger, the BPO company that he and a partner founded. There is a certain endearing boorishness about him as he "brags about how fabulously successful [Office Tiger] is." In short, the movie presents itself less as industry exposé and more like corporate propaganda for Office Tiger itself. As Anita Gates (2006) writes in her *New York Times* review of the film, "Executives suggest that Office Tiger's secret is working its staff remarkably long hours, eliminating coffee and tea breaks, and instilling pride in the employees' work by periodically telling them that they're the best and the brightest and that this job is the gateway to a glorious financial future for them. . . . [Office Tiger] lies somewhere between a white-collar sweatshop and a religious cult. But that may be true of a lot of corporations." As an intertitle from the film's introductory segment states, "Fusing American corporate culture and Indian ambition, Office Tiger aims to shape a new generation of global professionals."

Let us note some general characteristics of the "outsourcing industry" important for the analysis of digital media as an embodied phenomenon. India's software services outsourcing industry is a prime example of the globalization of affective or communicative labor. Over the past twenty years, the industry has grown rapidly and made significant inroads into the global market because of long-standing liberalization policies, the push toward technological modernization, significant state support, and cheaper labor costs in India (Balakrishnan 2006; Heeks 1996; Parthasarathy 2005; Upadhya 2008; Upadhya 2009, 9). As is well known, India's domestic manufacturers successfully negotiated the coming of liberalization in 1991. In the subsequent years, Indian corporations have not only kept competing imports at a low level, but have also begun to export on a larger scale than before in medium- to high-tech (i.e., high-value-added) areas. Forging a hybrid but compatible path from deregulated neoliberalism and state-controlled planned economies, Indian economic policy has enabled the general expansion of indigenous corporations, privileging the big over

FIGURE 2.1. Business process communicative labor. Still from
Office Tigers (2006).

the small enterprises. Over the past few years, major companies such as
Tata, Reliance, Infosys, and Wipro have floated their shares in Western
stock exchanges and acquired some well-known firms. "However," notes
Nirmal Kumar Chandra, "India's major breakthrough has been in information technology (IT) and IT-related services like software development,
'business process outsourcing,' etc. In the 1990s, Indians capitalized on
low labour costs here to seize opportunities that opened up with the IT
revolution in the US. Over the years the established firms and start-ups
moved into ever more complex areas of software engineering" (Chandra
2009). Postliberalization saw the emergence of qualitatively and quantita-
tively new information and communication technologies that have transformed many parts of the country (mostly in established and emerging
urban centers) into places where digital media, distributed databases, and
algorithmic perception have rapidly come to dominate various forms of
life and work.

Crucial to this emergence has been the BPO industry. This industry was
viewed by many in India as one of the primary engines of the country's
development over the first decade of IT-driven growth in terms of GDP
growth, employment growth, and poverty alleviation (Kuruvilla and Ranganathan 2008). India is the world's leading offshore destination for BPO
work, accounting for over 65 percent of the total value of work that is off-
shored in services globally. In the early 2000s, the outsourcing industry in

India was growing at around 40 percent to 50 percent per year, generating total earnings of US$39.6 billion in 2006–2007, of which $31.4 billion were from exports. Thus, BPO networks have played an important role in India's integration with world markets and in its recent economic performance.[6]

India produces the largest number of English-speaking graduates in the world, particularly engineering graduates—an outcome of Nehruvian state-led development policies and the country's extensive public and private higher education system. Of course, the wages of these graduates are considerably lower in comparison to overdeveloped countries of the global North, thus giving postcolonial India a "competitive advantage" in the BPO industry. India's 1.5 million IT professionals are engaged in different forms of work, from software design and development, coding and testing, to back-office operations, working for clients located primarily in the advanced neoliberal economies (especially the US). This fast-growing and creative multitude of IT workers is the BPO industry's most important "resource," and "controlling and coordinating their time, labor and knowledge is a critical task for managements" (Kuruvilla and Ranganathan 2008, 43). Adding to this human resource advantage is India's time zone (10.5 hours ahead of New York), which gives it flexibility in working hours—although, as we shall see, this fact is abused by management to extract more work for less pay from BPO workers. In the view of North American and European companies, all these factors make the Indian BPO company more efficient and cost-effective. Not surprisingly, the rapid growth and entry of new BPO organizations in India have resulted in the poaching of employees in high numbers. Contributing to this force of attrition, over the past ten years, salaries at the entry level have risen 10–15 percent while the increase is 25–30 percent at the top level (Anantharaja 2009; Kuruvilla and Ranganathan 2008).

The Indian BPO industry provides a robust case through which to explore the politics of digital networks and specifically the emergence of a kind of digital perception. First, as Carol Upadhya notes, much of the theorization on the "digitally networked workplace" is based on studies in the West, and India's BPO industry provides a wider base for comparative study. Second, this industry is characterized by new forms of 24/7 online labor, dispersed teams of knowledge workers, networked and modular production, and the constant digital surveillance of productivity, all of which is central to the global information economy (Hakken 2000; Upadhya 2009).

Let us highlight some key aspects of BPO work that enable us to focus on the implications of how the embodied experience of affective labor is tied to a new image of communication, creativity, innovation, and time (what I will call the jugaad image in chapter 3). First, let us keep in mind that BPO is, strictly speaking, itself a workaround: it is working around relatively more regulated labor markets, domestic workplace politics and norms, and national regimes of accumulation in the global North geared toward maintaining the welfare state. So for firms to work around these constraints to capitalist valorization, new strategies of both management and logistics (as we will see, the two are becoming increasingly indistinguishable) had to be put in place as a global solution to a threatening accumulation crisis. This was better known as the neoliberal globalization of the "free" market, and its heyday was in the late 1990s and early 2000s. Second, in India, as Upadhya and Aneesh both argue, while the bodies of Indian software workers were becoming more "immobile" (in contrast to the physical mobility of the "bodyshopping" or educated immigrant labor system of the 1980s), their mental labor was "mobilized" or "liquefied" (Aneesh 2006), "flowing through computer and satellite links as they collaborate and communicate with colleagues, customers and other teams located on the other side of the globe" (Upadhya 2009, 4). Third, this labor is managed through a postcolonial and hybrid ideology of immaterial labor, or knowledge work itself. The separation of intellectual and manual labor that we highlighted in the introduction by drawing on Alfred Sohn-Rethel's study is important to keep in mind here. In postcolonial India, the modernizing state celebrated gendered images of honest manual laborers as India's primary heroes of progress (consider the socialist imagescape of the cinematic classics *Shree 420* [1955], *Mother India* [1957], and *Purab aur Paschim* [1970]). Yet this was a technocratic socialist image that always existed in political and material tension with class, gender, caste, and regional hierarchies throughout India, and which have historically privileged intellectual over manual labor. Those "innovators" who have developed strategies to work around or hack into these hierarchies, liberating quanta of freedom and creativity, become the new heroes of smart-city urban chic/geek.

The contours of today's managerial image of knowledge networks is clearly apparent in *Office Tigers*: throughout the film we see how through video surveillance, team competition, networked innovation, and constant training that "control over the knowledge of employees is a key objective" (Upadhya 2009, 6; Yang et al. 2002). This managerial image of affective labor is a kind of cultural labor process that emphasizes "communication, collaboration, teamwork and knowledge sharing through building strong social networks" (Hakken 2000; Upadhya 2009). However, in India, as in many other places around the globe, crucial to the control of communicative labor is mobile phone–enabled direct monitoring of workers' activities, and elaborate information management systems, both of which allow management to extract ever more surplus value from workers, something that managers consider carefully when developing their information-auditing protocols.

Notes from the Smart City's Undercommons II

This parataxis attempts to bring the consideration of business process outsourcing into historical relationship to the dynamic field of everyday jugaads that are continually and increasingly blurring the distinctions between work and life today. Indeed, as we shall see, jugaad ecologies preceded business process outsourcing, and became part of their informal infrastructures and technologies. Consider the jugaadu Renu Rajawat, age twenty-three (interviewed by Anisha Saigal on September 14, 2015). She has lived in New Delhi all her life, she is a sociology graduate, music aficionado, and a "terrible drummer." Currently, she works as a freelance music/artist manager. Her parents have been residents of Delhi for more than sixty-one years. However, her ancestors came from Jammu and Himachal Pradesh. Her formal work has an informal ecology: creative support services are "completely and absolutely of a freelance nature. I work out of my room, terrace (making full use of my neighbor's Wi-Fi when mine isn't working), city cafés, and public parks."

Renu's work continually traverses the virtual, digital, and material physical planes, surfing a mutating multiphylum:

> I'm responsible for artist and event management, handling tour management; building strategies to ensure accelerated growth of the artist and any/every kind of promotional deeds. . . . I love it! I'm extremely passionate about the music/art/cultural space and I quit my job last year in October to do all things unimaginable, things I've only dreamed of. However, I do think that I will need to take up other avenues to be able to manage my expenses "regularly." No, I don't always have to travel to

work. I'm a freelancer and this gives me the liberty to schedule my work meetings either virtually via Skype or centrally (and closer) within the city. . . . I used to make use of every form of public transport until last month—now I drive almost every day. But it depends merely on the distance I have to take; I switch over to the Delhi Metro if my travel is NCR [National Capitol Region] based. . . . The Google Maps app is my best friend for my daily travels because it alarms me of traffic/construction settings. Another old school alternative to this is radio. Radio jockeys in most Indian cities give regular updates on traffic situations.

Renu uses both analogue electronic as well as digital/mobile media. "I religiously follow the new English [high-definition] channels on TV. My use of radio is only restricted when I'm driving around the city. Otherwise, my daily dose of music listening happens via my iPod/internet. I also follow a lot of indie (music and film) blogs, music magazines and eponymous photo sharing webs like Pinterest." Renu still likes to read actual printed books but also watches movies, listens to music, and reads PDFs on her laptop, computer, and phone. "Mobile phone for all the above mentioned, except for books. I can never abandon the print; whether it's the books or newspapers. My nonworking weekends go into reading books and exploring new music courtesy of Indieshuffle and 8tracks—app-based! . . . However, I do prefer using my laptop for heavy consumption webs, webs that don't yet offer applications. I'm also a strict laptop user when it comes to drafting documents/Excels. . . . It is much more comfortable."

Renu owns "an Android—Nexus 5. I bought it off Flipkart using my laptop because my old phone (a BlackBerry) won't support the app. Sounds as ancient as Akbar, I know. But Android really is the torrent of the telecom industry. It is uncomplicated, accessible, and easily exploitable." She does not share her phone at all, but "this in no way means that my phone lugs all things explicit. I'm just too particular about my privacy." Expressing an acute consciousness of her generation's connection to cross-platform technoperceptual assemblages, she declares with confidence, "Our generation and the one following us is considerably familiar with the controls of anything that is to do with technology. However, I did learn a few hacks from my brother to restore the sanctity of my phone. Hacks like battery-saver apps, ad blockers, and sound boosters. Did you know you could capture photos while shooting a video? Yes! These are essential travel hacks."

For work, life, and pleasure, Renu is connected to her mobile. So battery life is very important: after extended "heavy" use, "my phone barely survives a five- to six-hour clock. I have to keep a portable charger handy." Note, then, that the jugaad device so heralded by contemporary digital media, management gurus, and smart-city policy wonks itself needs a jugaad to keep running, given India's unstable energy grid and extreme weather conditions! Still, Renu "loves" her mobile. "This phone has been at the cutting edge of the market. I love the processor and its speed; the phone came with the Android 4.4 Kitkat, and the camera, much to my surprise, is remarkable. Oh and we were first in line for the Lolipop [the latest Android operating system] update—so yay!" Yet such digital commodities cannot escape the disappointment endemic to commodity culture more generally: "Being a music enthusiast, I was extremely let down with the sound 'levels' of the phone. It was almost nonexistent! The phone's battery is not removable, which for us forever tech-panicked Indians is a wicked thing!" Renu's use of the phone is largely focused around "making 'normal' calls for sure—that's why it's a phone! There might be a counterargument that app-based calls can also be made but most of them require Wi-Fi and have terrible connectivity. At least in India. Other than that, I use it for texting [Whatsapp], exploiting the very web of Facebook and photo sharing on Instagram. . . . [I use the mobile] throughout the day, for most reasons. Be it regular email checks, use of social media to stay in touch with friends, family or to stalk really, really famous people; and also for usual internet playtime, reminders, alarms, food apps, calculator, and, of course, the camera. Also, I am a huge fan of gifs, whether comic or otherwise. Since they're hilarious and save video buffering time, a three-second gif could possibly be the highlight of my day."

But like everything in a jugaad ecology Renu's mobile encounters its own feedbacked plasticity (both virtual and actual affordances relating perception to mobile) in her varied practices of jugaad/repurposing technology: "I use my phone as a mere torch [flashlight]; sometimes I scan my documents—especially my passport because it makes it look sharper than most scan machines. I have also used my phone as my laptop's mouse once and I use it to dodge Delhi's mental traffic on a day-to-day basis." This affective plasticity is lived for Renu as the relative "openness" and affordability of digitally networked communication technologies. "Traditional media are a closed system, whereas the mobile/digital media is much more open, far less expensive, and 'mobile.'" For Renu, people are inclined to the mobile or digital media because it provides connectivity in an interactive and, even, personal level. It is also significantly useful, such as creative apps engaging the interests of a user. Mobile media represent a form of rapid technology and accessibility. On the other hand, traditional media are more "actionable," especially television and radio. There are great advances in television technology, with apps in smart TVs engaging a majority of the population. This remains one big similarity with mobile media and in fact, gives traditional media the upper edge as it interacts with the user better with the help of a larger screen.

For Renu, jugaad

is nothing but a set of clever fixes curated to manage and curb complications surrounding our day-to-day struggles. As simple as it may sound, it does require a particular skill set to achieve the unachievable, mend the otherwise damaged or do more with less. In fancy terms, it is called DIY. I'm sure jugaad is our [i.e., Indians'] brainchild. You'd want to know how/why? Here you go: [she shows the following website on her phone: http://www.scoopwhoop.com/humor/pic-of-jugaad/.]

I survive off of jugaad, just like how Voldemort lived off of the Unicorn's blood. I'm only joking, but, yes, a huge part of my field of work does require countless jugaads, especially during event management. Jugaad is often seen as uncouth but it really is a matter of excellence and flexibility of thought. Here are a couple of examples from my life—stating both the way and reason: (1) I've used my laptop/phone camera to get ready/comb my hair because a lot of my friends who put up in hostels lack a basic mirror. Also when driving—not while! (2) I have tried clothes in the mall and have come back to use a shopping app on my phone to order the same product at a discounted price. This is smart shopping but is also a very faint jugaad strategy for the middle class. (3) My last working place didn't have a coffee/water warmer, so I would just place my mug next or atop my laptop to keep it warm! (4) During college, I would often miss lectures as I was working in the indie industry, which was taking up a lot more of my time than anticipated. I would get jugaadu "medical certificates" made by my then uncle doctor and walk away easy. I'm sure every college student has done this! (5) Professionally, my team and I live off of subtle favors or, as they say, "jugaadu connections." My favorite one: I once traveled to a fancy music festival as a band's "ostensible" photographer. I didn't even own a camera then! So much for travel and music! (6) Did I mention to you that I use a "rooted" phone? Phone hacks 101!

Rooting her phone allows Renu to run the sudo command, and it gives her enhanced privileges, allowing it to run apps like Wireless Tether or SetCPU.[7]

As a "curator" of jugaad, Renu creatively reproduces her (and her family's) social being through an extensive hacking ecology of media, energy, perception, code, control, habituation, and commoning practices. Her set of habituations—millennial, networked, time-intensive, affective, and plastic— expresses the phase transitions afoot in jugaad cultures of neoliberal technoperceptual assemblages. Her confident life-hacking ethics show in practical ways how hacking ecologies traverse the informal and formal economy, blurring legality and illegality, indeed mutating perception toward a kind of repurposing to the nth degree, experimenting with the social encounter and collective assembly dimensions of urban life along intensive gradients in coevolving ecologies of sensation.

Time, Process, and Management in Neoliberal Affective Labor

Such jugaad ecologies allowed for the social and organizational contexts of business process outsourcing to develop its by-now quite well-known strategies of flexible accumulation, and in feedback loops to develop systems for managing these very ecologies of sensation. Note for Renu above, saving time, economizing on time, compressing time in the jugaad time of her creative curation of life hacks are important dimensions to her newly digitized perception and practice both of work and life. Drawing on her ethnographies of IT workers' daily routines in Bangalore, India, Upadhya gives some concrete examples that allow us to grasp what these managerial techniques mean at the level of embodied labor. There are, first, sophisticated centralized software tools and computer systems to track workflows, output, and progress, similar to those found in call centers. In the time management system (TMS), for instance, engineers must use the timesheet to record time spent each day on a series of specified activities such as coding, attending meetings, and reviewing code. A tool for billing the client, this log also measures internal gradients such as productivity and "effort variance." As Upadhya shows, here productivity is defined as the "percentage utilization of resources," and is calculated by dividing the actual hours worked by the standard eight hours (2009, 9). Collated into status reports, these data become an object of review for the quality assurance department, which then gives feedback to the teams and project managers on their performance. These control systems allow managements to continually, and more and more meticulously, monitor workflows and performance by benchmarking tasks completed against the project timeline. The data thus generated are also used for other functions such as making estimates for project bids. Upadhya notes that such time management systems in fact constitute only a small part of the overarching control mechanisms for quality assurance,

> which if followed in its entirety entails 38 different procedures and 212 forms to be filled out. This large amount of "paperwork" is one reason why most engineers resent having to follow "process." . . . The TMS [Time Management System] described here constitutes only a small part of the . . . quality assurance process. . . . Thus, far from "empowering" employees as they claim, Indian software services companies have adopted a range of exacting neoliberal management techniques in their quest to gain control over the software labour process. Although these organizations present an image of "open" and "flexible" workplaces, ultimately they must keep tight control over workflows in order to maintain their profit margins. (Upadhya 2009, 10)

Process, in this mode of direct and indirect management, has become a master word for inculcating and monitoring norms of "capability" throughout the outsourcing workforce, and it happens through the micro-logging of activity and thus the strict management of the workers' time (Mutch 2016; Shaviro 2014). BPO companies bill clients on the basis of "man-days," and projects operate under strict timelines rendering control over time central to control over the work process itself. Micrologging and strict timelines facilitate extracting the maximum amount of time and effort from software engineers. Speeding up and closely monitoring the flows through the network are the keys to profitability. The Indian software industry is known for the very long working hours that are put in by software engineers—typically ten hours or more per day. Notes Upadhya,

> A major reason for this pattern of overwork is that the man-days required for a project are routinely underestimated when making bids, to

FIGURE 2.2. Rewarding the "tigers." Still from *Office Tigers* (2006).

keep the cost estimate down. This forces engineers to work much longer than the stipulated eight hours per man-day in order to meet deadlines. Another reason is the time difference between India and the client site, which means that conference calls often take place late in the evening for the Indian team, when it is morning in the USA. Although in theory employees are allowed to come to the office later in the morning to compensate for staying late, they usually come in by 9:00 or 10.00 a.m. and still have to remain in office until the conference calls are over, until 8:00 or 9:00 p.m. (Upadhya 2009, 10)

My argument here is that such "time management systems" are precisely what jugaad practices of workaround and hacking displace in both the domains of image ecologies (in thought, memory, habit, and perception) and regimes of control (as instrument, target, and mode of different forms of power). Neoliberal time management systems are sensory-motor circuits involved in increasing the profitability of organizing time, information, and work in the digital domain. This is precisely why today in neoliberal India the blurring between the formal and informal economies happens in the jugaad time of hacking power, a durational and interstitial passage from one affective state to another.

I turn now to Mermin's documentary to explore the above hypothesis. In one scene, a white Jewish American management trainer lectures the

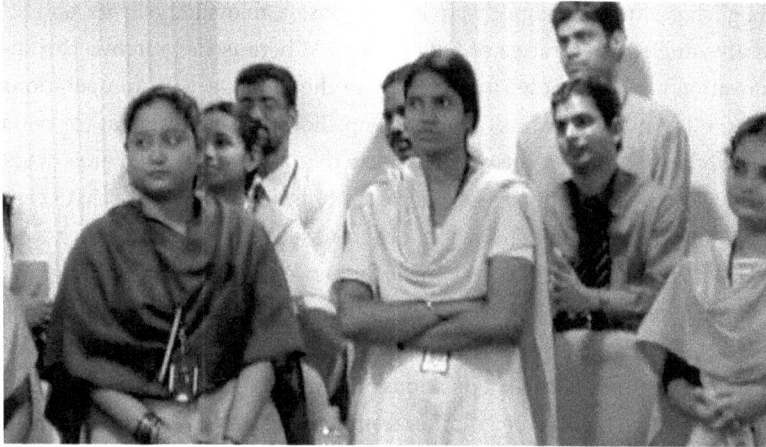

FIGURE 2.3. "Which team are you with?" Still from *Office Tigers* (2006).

"Talent Transformation Team." He tells a group of Indian employees that "a piece of history is taking place right now here at Office Tiger. It's not a call center that's content to do the simplest kind of work imaginable, make a profit and go on. No. Office Tiger is really thinking of innovative, much more efficient ways where the best and brightest of India can work with the best and brightest people all over the globe. You guys are truly a part of history. Such a rapid economic development in such a short period of time, such a dramatic change of values. But I have news for you: this is actually the first hint of what's going to happen, because this process of globalization is just beginning, and the opportunity for people like you—ambitious, young, talented people—is just starting." These words echo Richard Florida's influential thesis of the creative class catalyzing regional development in the overdeveloped global North; talent, technology and tolerance, the three Ts once proclaimed by Richard Florida as the key indices of a potential creative take off in any given region on earth (Florida 2012, 2017; see also Dubberly et al. 2017). Indeed, the early rise of bpo in India seemed a sure indication that a new creative class was emerging across urban metropolitan regions.

But *Office Tigers* unwittingly highlights the irony and brutality of this emergence. Intercutting nonmotivated shots of Chennai street life (Indian "truths"?) with interviews, the movie follows the English-speaking Indian employees through long, extended hours of meeting deadlines, learning

English grammar, dodging marriage proposals, managing clients, and literally singing the praises of Office Tiger. Where is the potential for innovation between these rituals? Early in the film, Deepak, an operations account manager, declares in a close-up shot, "I think it's great to spend twenty hours a day in the office because that tells you've got a great work ethic. I know I have done it in the past. I'm proud of it because that keeps me ahead in this competitive game. Because if I can spend twenty hours, you know, just being the best I can for those twenty hours, I know I've gained a lot of ground over all those hungry wolves around me." In contrast to Pritish's "personal" motivation for engaging in affective labor ("to meet new people"), Deepak positions communicative laborers as working for tactical advantage given an extreme aggressive, ever-changing field of labor relations; and yet we see these affective workers also operating in close-knit, sometimes transnational teams, where older, but still very much honored social and religious hierarchies shape interactions and work protocols. More, the film gives the false impression that women have an equal presence throughout all strata of Office Tiger; in that sense the ideology of meritocracy, or a worker's market value, works to neutralize the force relations of gender, caste, class, and religion of Indian society. Through obvious editing and camerawork, the movie makes clear that these implicitly masculinized and hungry wolves are not primarily other outsourcing companies but in fact Deepak's fellow employees, all of whom are organized into client-specific teams that compete for cash bonuses by constantly upping their own productivity. The bonus is for what is called sustained operations excellence, and the managers at Office Tiger declare their firm to be a meritocracy. Thus, hard work is correlated with merit; social networking is corporate networking and vice versa; innovation happens in the interstices of these intensifying work-becoming-life schedules.

What of the lived space in which Office Tiger functions? The spatial enclosure of Office Tiger establishes the business itself as a switchboard of connectivity, digital/video surveillance, and innovation incubation. The voice-overs usually are of activity inside the building, giving a sense of energy, productivity, and stability to the vast space. Later in the film, we see the high-security, executive offices of Office Tiger, located in what was, for a brief time, India's largest shopping center in Chennai. The securing of space through guards and closed-circuit TV, and the tracking of workers through scanned identification cards and cameras, not to mention the constant auditing of data transmission from each desktop computer, produce a rarefied space, distinct from the quasi-public mall (a privatized space for controlled public consumption). Overall, we can see the mode of attention, distraction and registration specific to these spaces: continuous worker registration flows into information streams of human resource data, and it is monitored through a variety of algorithm-driven practices, criteria, and sensors. Jugaad time in Microsoft Excel. Again, we see that the sensorium (or quotidian reality) is constituted by the cross-hatched vectors of surveillance gazes, modular cubicles, the constitution of subjectivity through specific forms of registration and identification, and the constant auditing of information flows, all happening in synergy with the processes of mall consumption.

So what is this form of labor that is being so closely modulated and continually disciplined? Aneesh argues in *Virtual Migration* that neoliberal economics has wrought a "fundamental transformation in the nature and organization of labor." As we have seen, with fast data-communication links, communicative laborers based in one part of the world are increasingly able to work with people from other locations around the globe (Aneesh 2006; De Angelis 2017; Hardt and Negri 2009; Peticca-Harris et al. 2015). Information and communication technologies such as the mobile phone and the personal computer increase management's ability to monitor the flow of work and the work of flow(ing) (Münster and Strümpell 2014; Terranova 2004). Indeed, the perpetual contact of digital networks, through webcams, login registration, global positioning systems, and so forth, enables management to automate the constant monitoring of employees who are away from their desks for work-related reasons. As employees are required to stay in touch with office staff from remote locations, making just-in-time discretionary decisions, the intensive engagement of employees is ratcheted up, resulting in an increase in effort per hour of work. Managers may view this kind of change in work organization as increased productivity; employees usually experience it as the squeezing of their capacities for private profit. As various studies have shown, "Workers are likely to experience more work, at an intense pace, under greater time pressure with more stress and heavier use of ICTs [information and communication technologies], as a single package" (Bittman et al. 2009, 687).

Keeping these contradictory transformations in mind, let me emphasize some of the key elements of the image of Indian services outsourcing in *Office Tigers*. Most obvious is the structured semantic slide between cultural values and capitalist value, which could be shorthanded as "the valorization of labor revaluates culture itself." *Office Tigers*, the movie and the organizational model, both elicits and covers over this dimension of cultural re-valuation through dichotomies of tradition versus globalization, India versus the West, the familial home versus the world office, presented again and again through familiar and emergent habits and patterns of communicative labor. This itself would complicate the too facile argument of the dissolution of boundaries and borders in the new transnational and largely informalized economy; what *Office Tigers* makes perceptible is the bodily implication, or affective regularities that distinguish value from value, population from population. Thus, in the form of the BPO corporation, the revaluation of culture is seen as necessary to the strategies of capitalist valorization, not only because culture changes the way workers think and act, but also because particular media ecologies inculcate new sets of resonant habituations correlated around the English language, information technologies, and market value.

In *Office Tigers*, the friction between the expectations of non–South Asian managers trained in neoliberal meritocracy and the network of social ties of its actual employees stages this disjunction of valuation. These ties surface in moments of guarded sociality within the office; in subtle looks of disgust or discomfort exchanged between managers and workers, and between the workers and the video camera; or in the opposition between traditional caste Hindu heterosexual kinship of "arranged marriages" and the "freedom" (for men) of the digital office. Moreover, the stark differences of status, wealth, language, and self-representation between the global services employees and the fleetingly visible populations that service them—the background labor of affective and material reproduction (care, maintenance, security, and administration staff)—bubble through this social disjunction.

Office Tiger provides measurable value-added services to firms mostly in the global North; indeed, this international division of digital labor, with its specific control regimes of relative surplus value, is based, as numerous commentators have noted, on the communicative labor offered through BPO work in the global South. This labor is always unevenly integrated with global supply chains and is constantly exceeding the boundaries between home and office or rural and urban, between merit and privilege, public and private, or work time and leisure time. This suggests that BPO labor is both potential (patterned but unpredictable innovation circuits) and habituated (image clichés and sensory-motor circuits). It is this space-time of both potentializing excess and actual habits that contemporary information and communication technology companies flaunt as the creative interactivity of information itself. Potential is a brand in the affective transitions of controlled creativity, in the in-between times that Office Tiger attempts to control as its own domain: "We liberate potential," as an intertitle reads. Indeed, it is the value of potentializing duration itself that is intensified by the value-capturing processes of Office Tiger. For instance, starting work on time, the duration of the work day, its lag from Western work time, the conscious deployment of mobile phones in the organizational structure facilitating the intensification of labor time through multitasking, employees' hoped-for future: the many dimensions of the neoliberal "time zone warp" (Aneesh 2001, 2).

The actual connectivity between work and digital control is maintained through an algorithm-based governance structure that Aneesh terms "algocratic." As Upadhya remarks in her review of Aneesh's study, the algocratic form of management depends on digital technologies, which structure work routines and workplace behavior. As we have seen, in India, in the postliberalization economy, many work tasks are now performed through digital technologies of different kinds and the symbolic manipulation of code, giving rise to new systems of control, based on that very coding process. "The algocratic mode has enabled new global flows of information labour as well as control over geographically dispersed workers through constant online access and monitoring, as seen in the model of geographically and temporally 'distributed development' followed by Indian software outsourcing companies" (Upadhya 2008, 345). Indeed, the digitization of information and its circulation in real time across the globe are key catalysts in the transformation of value (Toscano 2008). Michael Hardt (1999) notes the recursive nature of information and communication technologies through which distributed digital networks can continually modify their own operation through feedbacked and interactive use, such that even the most rudimentary forms of artificial intelligence allow the computer to modulate operation based on interaction with its user and environment.

To summarize, I have argued that there are specific capacities that are both potentialized and actualized in the value-added image of BPO labor; an example of which is the TMS. In this form of communicative labor, the BPO perception is instrument, target, and mode of power; as a sensory-motor circuit in the form of, for instance, the TMS the perceptual machines of algocratic management modulate directly networked labor. This emerging sensorium or quotidian reality brings together in one digital and physical space surveillance gazes, modular cubicles and hot desks, and the constitution of subjectivity through specific forms of registration and identification. In short, we see that the affective structures (rhythms, durations, temporalities, dispositions, and their material and intensive flows) that have developed in BPO modulate work through temporally conditioned habits and innovative strategies for capitalist valorization.

Conclusion: The Value of Affect

My method here focused on diagramming India's emergent sensorium in the forms of affect, habit, and control in BPO in India. I addressed several decisive factors in the emergence of digital cultures in India. First, because of its centrality in understanding the rituals of the everyday, the feminist question of the cyborg body within contemporary communicative forms of capitalist valorization—BPO—has been pursued by rejecting the mind-body dualisms still so central to much of contemporary studies of digital cultures. Instead, taking inspiration from the critical tradition of feminist materialism, I have attempted to diagram a new ontology of sensation in the domain of digital culture in India. More, the notion of process has been foregrounded in the method of diagramming social practices. I focused specifically on the organization of duration and temporal cycles in and through this emergent capitalist ecology. As we shall see in chapter 3, the jugaad image disturbs and reorganizes this ecology through practices of innovations that happen when time is less, not more minutely, parceled; recalling Nikhil's "light-bulb" expression from the introduction, jugaad events occur when distraction is allowed to enter into everyday life more centrally, and so they produce new connections and value-generating innovations. Finally, I suggested how diagrammatical method traces the connections and feedbacks between the processes that constitute a given multiplicity of perception, affect, bodies, and information to develop the common notions of their relations. To understand what is at stake in this method, it is important always to consider the emergent capacities and changing relations of force implied in the assembling of these processes, as it moves analysis to a feminist posthumanism of information and communication technologies. The ethical implications of this are that technoperceptual assemblages such as are found in BPO are not natural, ahistorical phenomena but rather fully historical and strategic machines for the creation and capture of value and affect. Moving these assemblages toward an affirmation of a global undercommons whose informal economies of commoning hack into capitalist, racist, and patriarchal power demands an ethical experimentation in the material and intensive composition of these assemblages themselves (Deleuze 1988b).

I also suggested that there are synergies forming in the temporal modulations of work and life under neoliberal habituation in India. Indeed, an emergent set of functional connections is taking shape between the economic, technological, and cultural processes of digitized labor in India's smart/Big Data cities. This emergent functionality is what gives India's technoperceptual assemblages their specific dynamism. The question remains: What are the kinds of affects (capacities) that resonate with the embedded timescales of embodied sensation, from micro-durations in specific regimes of passage such as the time management system in BPO (space-times of sensation) to population-specific becomings that form over decades and generations (slowly changing vision pathways in the brain)? The aim of this chapter was to examine the question of digital connectivity from the perspective of affective capacities and habituation. The implications of this come to the fore here in what *Office Tigers* offers as a diagram of contemporary information and communication technology's potential value in India: in the modulation of perception, attention, and memory on the one hand, and the networked mobility of digital services and their algocratic control on the other, the potential commoning of India's new digital reality is posed. It is toward a hacking ecology of this digital reality that we now turn.

/// FABLES OF THE REINVENTION II ///

New Desiring Machines

The Minor Events were inaugurated by a huge, unplanned explosion at the mutation camps in Bhopal during the festival season of that year, when Id and Diwali fell on the same day. The stars have aligned but like fuckers, chutiye jaise, they said immediately at the paanwalla's near Lyall Book Depot on Sultania Road. On that day, no Muslim fast, no Hindu Ramlila, but instead a becoming "mela" of different crowds, fleeing in panic through the Old Muslim City, where the Begums of Bhopal ruled for a century and then through the new Administrative Smart City (this was the networked center through which the Hindu fascists had converted Bhopal into the capital of the new Hindu Rashtra). The Muslim Old City and the Hindu Smart City: between and beyond these securitized spaces a hypernetworked interzone of jugaad practices brought the codes of security to the hackers of the Minor Events, some of whom lived and died just beyond the gates of Union Carbide's crumbling, deserted murder factory. Both spaces were marked by conflicting histories of conquest and colonization, mutating ecologies, enfolded. The networking of the two cities was part of the story of that phase transition. On that day, we were visiting relatives on newly garlanded scooters, calling and texting bent on efficiently coordi-

nating trips; placing orders; getting crisply ironed saris, kurtas, and shal-
wars; enjoying syrupy sweetmeats, ajwain-nourished bhajhiyan, and gin-
ger chai—the things it seemed that people had been doing for centuries, if
not millennia. Everything just going on, like that only, and then suddenly a
blast and an enormous plume of gray sootlike gas forming over the south
of the city, over Char Imli Hills, a moving cloud as if not on the wind,
brooding and inscrutable. The murderous cloud crept again over Bhopal,
this time from smart city to gated Muslim city. We watched the gas cloud
become volatile as it went through oxygenation, suddenly crackling. We
watched, mesmerized, and I remember thinking it will make people happy
even though I knew immediately that it would bring nothing like the relief
of the monsoon clouds. The cloud burst across the administered smart
city, generating a kind of internal combustion and moving past the Jamma
Masjid on the north side of the city. And as it moved through different
bodies, depending on how imbricated their ecologies were with the water
supply in the municipal area, bodies began to "reinvent" (an IIT-Bhopal
adviser to the administration would later come up with the brandable
term) new silicon- and carbon-based assemblages. Later that afternoon,
the first reinvention captured in data thrilled bystanders at the paanwala's
near Lyall Book Depot, some thinking its explosion was a trick wrought
by the Bollywood types (that's what the Bhopalis say). That first captured
Minor Event, when rendered as a humanly perceptible audiovisual stream,
seemed to show a group of people, all mumbling in the same undecipher-
able drone, moving deliberately down Sultanya Marg, toward Munna's
paan shop. In the audiovisual rendering of the captured data, the mutations of the body expressed themselves on the skin, which would start
to emit electricity in short bursts, on video it looked liked they were on
fire, but they weren't. Other data renders indicated that they were exhal-
ing pure oxygen, their bodies were accelerating all sensory functions, and
their bodies were conducting information and burning their clothes. That
Event went viral across the multiphylum: how could anyone call them "minor" after that?

Many people across Bhopal died immediately, tens of thousands: those
who could afford it fled, and just as many who were directly affected by the
cloud were "encamped." But the mutations dominated the headlines. As
Bhopalis had already suspected, from the film industry first they came to
Bhopal to see the "reinvented," to experience firsthand the newly mutating
chemical phylum. Creative treatments were planned; pitches were prom-

ised. People hailed it, saying this and that about this and that scripture and deity, blessings and glad tidings, information and karma, but something was actually moving through those people. It made one wonder what their bodies could do. But my research shows that there was nothing simply spontaneous in the Minor Events; they were the patterned but unpredictable emergence of new organizations of ecologies of practice.

Which practices? We affirm it even, especially today, on the brink as we are . . .

How can we write a History of the Minor Events without turning it into the occasion for another Lament of the Failed Revolution—and how else can we understand the Movement of Reinvention? Rather in order to invite another potentiality to inform our future, I want to know what was/is becoming in and through those ecologies of practices that seemed so central to the Minor Events. What in them has eluded, exceeded even their own history?

The grounds of the Union Carbide plant were located in the poorest parts of the Old City (by the train station). The gas that was emitted from the Taylorized plant (cost cutting led to eventually deadly workarounds in terms of workers' health, safety, hazardous materials storage, repair materials, and staff training) on the night of December 3, 1984, devastated majority Muslim and Dalit communities, killing around three thousand people immediately. Those grounds, not far from what had been the largest electrode plant in Asia, were never cleared of their toxic sludge, and over time, they entered the drinking water supply and permeated an iron-rich soil, already volatile due to the monoculture farming of soy for export, a newly networked chemical phylum that came to mutate our form of life.

But mutations claimed our becomings through other modes as well, and it was the confluence of mutational lines—a phase transition in the multiphylum—that prepared the way for the Minor Events. The Union Carbide plant produced, among other things, a gas—methyl isocyanate, which reacts readily with water to produce carbon dioxide, urea, and heat. This gas entered the groundwater and over time mutated (partly diminishing, but also accelerating) people's capacities to photosynthesize carbohydrates from carbon dioxide and water. It seemed to them living at the time when the mutation was first studied in any depth that breathing was itself *nourishing*, but we now know that this seeming nourishment withdrew the oxygen from the subject's ecology itself, and drastically shortened their lives.

And the heat of the new chemical phylum was also accelerating as it were. Globally the average temperature had rapidly increased, producing unprecedented ecological disasters, from the disappearance of entire island nations to the "reclamation" of Mumbai by the sea. Even before the Minor Events, temperatures in Bhopal would regularly hit 50 degrees Celsius. Global temperatures had become a tradable future in the formal and informal economies, and its relationship to ecologies of data and digital practices that had thickened interactions rapidly across India's vast subcontinent was actively speculated on. These rapidly changing environmental and financial conditions catalyzed specific coevolutions in technoperceptual assemblages, some nodes of which were in a peculiar state of readiness in Old Bhopal. During research into the possible coevolution of the mutations at the site of the Union Carbide plant with the Hindutva projects around artificial life, I found that in poor communities, and especially for women, girls, and elders, anyone who relied on the groundwater anywhere in Bhopal had an average mobile phone battery life of six months on one charge, as long as they stayed in tactile connection with the device for at least an hour a day. Gradually the chemically enhanced battery would burn up all the components due to the oxygen and heat generated by the mutation (the company's copyrighted innumerable devices resisted the "burn" of Reinvention). Communication, having already become readily 24/7 through a rigidly enforced registration-in-connectivity, now more intensely, more molecularly, neurologically attuned, entered into all the minor and major ruts of habit, deterritorializing their refrains, creating unpredictable frictions, sometimes even bottlenecks in information flows, enlivening the mutations with the artificial lives of the interwebs, with its spectacles, conspiracies, and commodities. People's attention were splintering and crystallizing rapidly; you saw them walk around in a daze, speaking as if to themselves, as my research clearly shows. Lost in quasi-individualized, quasi-collective bubbles of media (dis)interactions and consumption the mutations turned these habits into a veritable ontogenesis machine for imperceptible becomings. These were the far-from-equilibrium conditions of habituation that precipitated the Minor Events in Bhopal, and without resemblance or contiguity they resonated with different struggles across the postcapitalist world.

3

JUGAAD ECOLOGIES
OF SOCIAL REPRODUCTION

In this chapter, I pursue the diagrammatic method assembled through critical parataxes of and with intensive philosophy, cyberfeminism, and postcolonial and autonomist Marxism. The aim of this chapter is to bring the earlier conversations regarding jugaad practices into productive tension with the contemporary feminist and postcolonial critique of capitalist social reproduction (women's unpaid work) as an ongoing force of "primitive accumulation" (Marx 1976 [1867]).

Recall that empiricism, for Deleuze, developed the idea that relations are external to their terms (Bains 2006, 24–25; Blackman 2008; Deleuze 1991, 98–99). Recent studies have drawn out the social, economic, and political implications of this relational method for affective ethnographies of contemporary urban ecologies. Ecologies of social reproduction, the reproduction of the commodity labor power as an expression of the ecology itself, stage and control contradictions and antagonisms in its material practices through what Alfred Sohn-Rethel called the social synthesis, the network of processual relations by which society coheres (Sohn-Rethel 1978, 4). In Dalit ecologies, due to caste hierarchies, debt structures, vibrant village-city feedback loops, and other institutions of power and privilege (e.g., political parties and schools), the social synthesis is always a failed promise of universal productive citizenship. These subaltern ecologies of social reproduction pursue becomings in working around or hacking into entrenched power and privilege. As we saw in the last chapter, these "innovative" workaround practices can form the basis of a new extreme work ethos in neoliberal India: whatever it takes to get the job done.

Other studies have shown that a careful attention to the distribution of agency in ecological processes, visceral experience, urban metabolisms, and embodiment can yield important insights into the space-times of affect (the relational and temporal capacity to affect and be affected). Drawing attention to distributed agents, as dynamic and feedbacked elements of human and nonhuman, or technoperceptual assemblages within biocapitalism, scholars have shown how these dynamics self-organize in surprising, sometimes noncapitalist and disruptive ways (Arboleda 2015; Cameron and Gibson 2005; Clark 2003; DeLanda 2002; Gibson-Graham 1996, 2006, 2008; Gradin 2015; Mitchell 2008; Posteraro 2014; Safri 2015; Safri and Graham 2010; Thrift 2007). Hacking in digital ecologies is one example of such potentially disruptive practices of contemporary post-colonial urbanism. While prior research in this area has focused on hackers' masculinist, visionary, and heroic identity formation in media practices; recoding digital computational networks of information; hacking for open-source, resilient urbanism; or "life hacking" for generating associations, this chapter brings hacking practice into fruitful conversation with critiques of social reproduction in urban India to affirm a form of biopolitical production that remains only unevenly integrated into capitalist relations of production (Amin 2009; Bannerji 2000; Coleman and Golub 2008; Jiménez 2014; Kubitschko 2015; Leyshon 2003; Routledge 2008; Wark 2012).

In this ecologism of economic and political relations, the sheer complexity and singularity of contextual and affective practices play out the shared human and nonhuman ontologies (Simpson 2013). There are important political and methodological implications to these questions into the space-times of affective environments of practice, many of which are beyond the purview of this chapter. Here, I aim to establish effective relations between geographies of urban political ecologies research and the feminist critique of social reproduction. I focus critical attention on the complexity and singularity of technoperceptual assemblages in affective processes of paid and unpaid social reproduction that form surprising and volatile ecologies today in urban North India. Taking the postcolonial critique of the *ghar*/*bahir* (home/outside) binary in the nationalist resolution to the "women's question" as a reference point, I focus on the affective and metabolic dimensions of women's domestic labor in emergent and residual social reproduction regimes. Developing embodied rhythm analyses of domestic work and media consumption practices, this chapter engages different relations of force, production, scale, and invention in specific biopolitical contexts affirming other ethical diagrams of their becomings (Arboleda 2015; Borsch et al. 2015; Burton 1997; Deleuze and Guattari 1987; Fraser 2008; Guattari 2005; Lefebvre 1991, 2004; Massey 2004, 2005; Prasad 2015; Sage et al. 2015; Weeks 2011). The argument is structured through an initial section that articulates together mobile technologies, jugaad and hacking ecologies, and biopolitical production in urban North India. I then draw out key linkages to and engage with strategic tools in the feminist social reproduction research that effectively blurs the (post)colonial binary of ghar/bahir in an urban metabolics of imbroglios (Birkinshaw 2016; Sage et al. 2015, 497; Swyngedouw 2006, 119); the final section is taken up with a critical consideration of a set of interviews with women involved in hacking ecologies in South Delhi. In conclusion, I further develop *Jugaad Time*'s diagrammatic propositions relating affect and hacking ecologies to the metabolisms of social reproduction.

Reframing Biopolitical Production

In recent analyses of technoperceptual geographies, such as mobile phone ecologies, a new frontier of capitalist expansion has been highlighted, one "centred on digital technologies, new divisions of labour, and an intensifying role of affective and mental assets in commodity production," profoundly transforming the rhythms and repetitions of "practical activity— and hence the way in which urban environments are produced" (Arboleda 2015, 38; Birtchnell 2011; Coleman and Golub 2008; Gibson-Graham 2008; Glück 2015; Kumar and Bhaduri 2014; Lazzarato 2006; Scott 2011; Starosta 2012; Vercellone 2007; Wyly 2012). In these studies, the subaltern "multitude" is related to processes of governance and legitimation in contemporary capitalist societies (Deleuze and Guattari 1987; Gandy 2006, 509). From the perspective of the subordinated multitude, urban environments are actively produced to work around or hack into power and privilege, the electrical grid, copyrighted mediascapes, the water supply, caste and religious institutions, urban flows of traffic, spatial and social divisions, and stratified identities. With the mobile phone, and with different styles and norms of technoperceptual subjectivity, in which practical activity is intensified, accelerated, and digitally registered through patterned but unpredictable interaction and exchange, multiscalar, hybrid carbon- and silicon-fashioned ecologies become central to what Michael Hardt and Antonio Negri have called biopolitical production (2001, 2004, 2009; see also Arboleda 2015; Besserud et al. 2013; Kubitschko 2015; Last 2012; Lefebvre 1991, 2003; MacKinnon 2011; Sage et al. 2015; Swyngedouw 2006; Tronti 2005). The forms of classed, caste, and gendered "pack multiplicities," shaped through culturally specific and often nonnormative practices of urban assembly and encounter, are both increasingly common and far-flung thanks to new mobile digital technologies (Deleuze and Guattari 1987). In this framework, the accelerating space-time of the city, as Martin Arboleda notes, enables specific metabolic exchanges—material and semiotic—that result not only in the transformation of urban space, but also in the production and transformation of social subjects (Arboleda 2015, 40; see also Ansems de Vries and Rosenow 2015; Lefebvre 2003; Merrifield 2013). Following Hardt and Negri, Arboleda suggests that the production of subjectivity is not enclosed within factory walls—as it was during Fordism—but has become diffused throughout the entire metropolitan territory, "as labour has become flexible and mobile, production increasingly larger scale, and the integration of labour processes more complex than ever before" (Arboleda 2015, 39; Hardt and Negri 2001, 2009). The argument of this chapter is that social reproduction in "smart" urban metabolisms—understood here as intensively involving organisms' "interaction with their environments and with the regulatory processes that govern such patterns of interaction" (Arboleda 2015, 37; Swyngedouw 2006)—actively participates in this new form, suggesting a speculative diagram of the multitude commoning (Berlant 2016; De Angelis 2017; Kubitschko 2015; Lamarca 2014; Harney and Moten 2013).

As we saw in the previous chapters, the mobile phone and its insinuating ecologies of rhythms rooted in patterns of work, information, bodily movement, pleasure, consumption, and caste, gender, and class reproduction frame our diagrammatic understanding of the distributed agency transforming postcolonial practice in urban India today. These emergent practices call for a reengagement with Henri Lefebvre's analysis of the social geographies of rhythms of repetition, labor, and sociality, and with its critique. Noting the "forceful or insinuating manner" in which a class imprints a rhythm on an era, Lefebvre argues that the politics of these rhythms directly affect social change (2004, 14; see also Lefebvre 1992). Further, pursuing Lefebvre's intuition, here in the rhythms of capitalist circulation and the production of time-spaces, we find the machinic assemblage of a certain biopower operating in the habituated repetitions and differences of logistics, communication, pleasure, and media (Arboleda 2015; Deleuze and Guattari 1987; Manning and Massumi 2014; Marx 1992; Massumi 2015a; Wood and Rünger 2016). Lefebvre (2004, 8) recognized how rhythms impose their structures, "truths," and "norms" in and on the body—through "respiration, the heart, hunger, thirst." He suggested (2004) that a disembodied grasp of rhythms is impossible; rather, it implies a close understanding of the repetitions constituting the laboring body under cognitive and postcolonial capital (see also Ajana 2013; Arboleda 2015; Ash 2013; Berardi 2008).

Yet, as Paul Simpson (2008) remarks, in "Lefebvre's Rythmanalytical Project we are mostly just looking at the body and how it is being acted upon by societal forces, rather than considering the visceral, elusory nature of the body itself" (824). Similarly, Edensor (2010) noted there is a need to conceptualize the body's capacity to affect and be affected by rhythms of repetition that "circulate in and outside the body [and] also draw attention to the corporeal capacities to sense rhythm, sensations that organise the subjective and cultural experience of place" (5); while Waitt and others (2013) argue for a visceral analysis involving a "move towards a radically relational view of the world, in which structural modes of critique are brought together with an appreciation of chaotic, unstructured ways in which bodily intensities unfold in the production of everyday life" (283–284). The notion of biopolitical production returns us to Elizabeth Povinelli's concept of geontopower; while mostly following Michel Foucault's critical project in assimilating geontopower (which is beyond the dichotomy of life and nonlife) to forms of domination (2016, 208–209, 240–242), recent attention to postcolonial informal economies that traverse the sacred and the profane, rural and urban, and stratified subjectivities suggests a different ontological understanding of both biopower and geontopower (Massumi 2014).

In the contexts of contemporary urban Delhi, this "biopolitical production" in relation to specific forms of affective labor is tied to mobile phone technologies and networks in various ways. Recent shifts in the political landscape for smart city agendas, such as the Aam Aadmi Party's free Wi-Fi initiatives, have sought to entrepreneurialize these networks in what Faiz Ullah and Aasim Khan have argued is a form of populism dominant throughout much of India today, but especially in Delhi (Khan and Ullah 2018). There, research poses the crucial issue of labor rights and working conditions in relationship to the proliferation of digital technologies. These policies come to capture what is already a vast and largely autonomous field of work(around) practices. For instance, communication patterns do affective work: maintain and keep active relations, express desires, consume and create media, coordinate movement (encounter, assembly) through urban space, and reproduce and blur the public/private distinctions within postcolonial Indian urban space. As such, the information revolution in India has been involved both in the discourse and practice of national development and in the "redefining and rejuvenating" of "manufacturing processes—through the integration, for example, of information networks within industrial processes" (Hardt 1999, qtd. in Arboleda 2015, 38; see also Birkinshaw 2016; Dean 2012; Moore 2015). In these postcolonial urban contexts, the infrastructures of capitalist production and circulation have their own pirate kingdoms and informalized logistics in urban metabolic imbroglios (Birkinshaw 2016; Liang 2009; Sundaram 2009). While most piracy networks are deeply imbricated in the value form of capitalism and hence the social relations necessary for its reproduction, these networks, expressing the complex social and economic antagonisms inflecting Indian urban ecologies, are constituted by processes of hacking and workarounds that are not reducible to, and indeed continually exceed, capitalist relations of production (Cameron and Gibson 2005; Gibson-Graham 2008; Gradin 2015; Jiménez 2014; Kubitschko 2015; Leyshon 2003; Routledge 2008; Safri 2015; Wark 2012). These hacking ecologies are also central to the implosion of India's jugaad state in recent decades, visible in positional conflicts between Dalit subcastes, between Dalits and Adivasis ("tribals"), between "Other Backward Classes" and Dalits, and between all of these social groups and Muslims, and the customary and long-established hierarchies of which are being restructured and sometimes challenged in and through new communicative and social practices in digital technologies (Birkinshaw 2016, 59; Gudavarthy 2016; Kaviraj 2010; Kumar and Bhaduri 2014).

These hacking ecologies have a common basis in the multitude's bottom-up biopolitical production—"languages, images, codes, habits, affects, and practices"—which "runs through the metropolitan territory and constitutes the very fabric of the modern city" (Hardt and Negri 2009, 250; qtd. in Arboleda 2015, 35; see also Adams 2014; Braun 2014; Ekers and Loftus 2012; Gandy 2006; Kraftl 2014; Braun and Wakefield 2014). Opting for the notion of "metabolic urbanisation," Arboleda offers a strategic view of the city as the "factory for the production of subjectivity" (2015, 38; see also Caffentzis 2005; Camfield 2007; Starosta 2012). Increasingly, a key mechanism of this production is the assemblage between mobile and networked technologies and differently situated (non)human perception through which technologies migrate and come to be contextually repurposed, accepted, modified, and operated. Research into technology usage in the global South has focused on technology ecologies of microentrepreneurs, and entertainment, attempting to reimagine the diagram of ubiquitous computing and technologies for populations in low-income, digitally unstable, and diversely literate environments (Rangaswamy and Sambasivan 2011, 553–554). As we shall see, in the context of postcolonial Delhi, the politics of caste, class, religion, gender, sexuality, and environmental conditions shape access to and engagement with digital technologies of biopolitical production (e.g., the mobile phone) across subaltern and neoliberal zones of contact and exception (Agamben 2005; Grover 2009; Pratt, A. 2009; Rangaswamy and Sambasivan 2011; Sundaram 2009; Webb 2013; see also Heynen et al. 2006; Holifield 2009; Kaika 2005; Loftus 2012; Swyngedouw 2006).[1] Indeed, dowry-related murders and violent controls and regulations on mobility and sexuality by family, caste institutions, and the neoliberal, authoritarian state are continuing and in some sense intensifying dynamics of social exclusion in contemporary India (Mandal 2013, 126; see also Radjou et al. 2012).

The Social Reproduction of Ghar/Bahir (Home/World)

I now turn to the question of the feminist Marxist critique of capitalist productivity and social reproduction. As economic geographers Katherine Gibson and Julie Graham point out, feminist economic analyses have demonstrated that nonmarket transactions and unpaid household work (both by definition noncapitalist) constitute 30–50 percent of economic activity in both rich and poor countries (2008, 615). Drawing on work by Jennifer Pierce (1995), Arlie Hochschild (1983), and Hardt and Negri (2004), Arboleda notes how characteristics usually associated with "women's work," such as family, care, or domestic work, which are entirely immersed in the corporeal, have become increasingly appropriated by capital and extended to all sorts of labor (2015, 38). Thus, both the category of domestic work as social reproduction and the space-time of domesticity are sites of feminist struggle that extend well beyond the confines of the domestic space (Weeks 2011). Socialist feminists in the 1960s and 1970s, in India and elsewhere, posed the problem of this site of the domestic as a question of the patriarchal reproduction of not only human capital but also gender itself. In this critical discourse, the development of alternative economic models of value and exchange is rooted in materialist analyses of social reproduction (Barbagallo 2015; Barbagallo and Federici 2012; Dalla Costa 2015; Federici 2003; Gopal 2013, 235; Laslett and Brenner 1989; Menon 2011, 18–20; Menon and Nigam 2007).

While socialist feminists using the concept of social reproduction in the 1960s and 1970s struggled to ensure the recognition of women's invisible labors that contributed to the enhancement of economic and social life, contemporary struggles to recognize women's labors have spread to broader patriarchal structures, informal logistics, scalar politics, and urban metabolisms (Arboleda 2015; Blackburn 2014; Cameron and Gibson 2005; Gibson-Graham 2008; Gradin 2015; Hardt and Negri 2004, 2009; Hepworth 2014; Latour 2005; Lefebvre 2003; MacKinnon 2011; Massey 2005; Safri 2015; Safri and Graham 2010; Sage et al. 2015; Swyngedouw 2006). For Camille Barbagallo and Silvia Federici, reproductive work refers "to the complex of activities and services that reproduce human beings as well as the commodity labor power, starting with child-care, housework, sex work and elder care, both in the form of waged and unwaged labour" (2012, 1–2). In this context capitalist productivity is based on what Karl Marx called a prior but constantly reproduced "primitive accumulation"—in this context, the accumulation of value through (overwhelmingly) women's (mostly) unpaid and often noncapitalist labor (Deleuze and Guattari 1983, 231–232; Federici 2003, 63; Gibson-Graham 2006, 2008; Marx 1976 [1867]). Barbagallo and Federici urge us to a renewed examination of the historically recent neoliberal restructuring of the global economy in its reshaping the organization of social reproduction, and transforming bodies and desires, kinship, workplaces, domestic space, and social relations. These feminist Marxists pose the problem of the intensification and institutionalization of the global economic crisis by composing alternatives "to life under capitalism, beginning with the construction of more autonomous forms of social reproduction, for every day it becomes evident that neither the state nor the market can guarantee our survival" (2012, 1–2; see also Beneria 1979, 1982; Beneria and Sen 1981; Bourdieu 1976; Engels 1940; Gibson-Graham 2008; Laslett and Brenner 1989; Sharma 1986).

The feminist critique of social reproduction in India has been shaped by (post)colonial politics and economy (Bannerji 2000; Beneria 1982; Beneria and Sen 1981; Ghosal 2005; Menon 2011; Menon and Nigam 2007; Sinha 2000). As Partha Chatterjee noted in his postcolonial critique of the nationalist resolution to the woman's question from early in the nineteenth century, British colonial discourse assumed a position of sympathy with the unfree and oppressed womanhood of India, thereby transforming this figure of the Indian woman into "a sign of the inherently oppressive and unfree nature of the entire cultural tradition of a country" (Chatterjee 1989, 622; 1990, 233–234; Tagore 2002). This created a gendered and spatial binary between a Westernized and Indian masculinist outside, and an essentially Indian domestic space of pure femininity within anticolonial nationalist movements throughout South Asia. As the ironic nationalist "bigot" Gora says early on in Rabindranath Tagore's famous novel (1909), "I affirm that all the exaggerated language about women that you find in the English books has at bottom merely desire. The altar at which Woman may be truly worshipped is her place as Mother, the seat of the pure, right-minded Lady of the House" (2002: 11–12).

While Chatterjee's analysis remained within representational critiques of ideology, the ontological dimensions of this lived dichotomy are carefully attended to in the analyses cited above. Today, in most parts of India (and elsewhere globally), women are commonly assumed to be "unproductive" members of a household; a well-known Hindi "kahavat," or proverb, throughout North India renders the domestic mode of production as "to raise a daughter is like watering a flower in someone else's garden."[2] The reproduction of these real and imagined hierarchies is tied to the changing historical importance and material relations of caste, class, and sexuality in India, and the question of discursive and material practices of "national development" and gender power (Paltasingh and Lingam 2014, 46). In their review of feminist social reproduction debates in India, Tattwamasi Paltasingh and Lakshmi Lingam suggest that dynamics of social reproduction remain critical to understanding women's historical class, caste, and sexual positions within the domestic mode of production; social reproduction also highlights the gendered distinctions necessary for the stability of not only domesticity, but also the reproduction of social life under capitalism (Paltasingh and Lingam 2014). They show how in feminist analyses understanding the complex situation of women in India—its unique features of stratification—opens research onto, for instance, the agrarian structure, caste hierarchies, the private control of property, and the adoption and adaptation of different technologies within domestic labor (Paltasingh and Lingam 2014, 51; see also Menon 2011; Prasad 2015; Tambe 2000).

Hacking Ecologies of Social Reproduction in Delhi

Recent work on the role of information technologies of ubiquitous computing in Indian urban ecologies has brought out the complexity of its "metabolic imbroglios" (Swyngedouw, 2006, 119, qtd. in Sage et al. 2015, 497; see also Anand 2011; Birkinshaw 2016; Jeffrey 2010; Rangaswamy and Sambasivan 2011). Consider for instance the logistics of water acquisition in Delhi. Matt Birkinshaw (2016) notes that exceptions to the formal water network quickly become fixed into the fabric of the city: pipes were laid multiple times with different connections created each time, either by installing an extra pipe for a new household or through bypassing a series of connections in order to improve their pressure. Thus, urban metabolisms such as water-pipe networks are repeatedly modified by residents and businesses. Birkinshaw remarks how the physical complexity in one small lane or neighborhood can be "daunting, buried underground, and the information is dispersed among residents, business owners, whoever has done the work, municipal and private sector staff. Information is not only fragmented, but guarded as people have endeavoured to provide themselves the best available service and do not trust external actors to disturb their set-up" (2016, 58).

AN URBAN METABOLISM can be characterized through its autonomous "bypasses," "workarounds," and "fixes"—all potential or actual metabolic imbroglios. These life hacks over time become stabilized and functionally accepted as part of the network, which is in any case of such complexity that "the 'official' network may only be known in certain areas to local engineers, and in its totality may not be known at all" (Birkinshaw 2016, 58).

Birkinshaw suggests that the widespread presence and power of informal interventions, both physical and administrative, have led to an "informalisation of the state" or a "jugaad [workaround] state" (2016, 59). In many urban ecologies in India, hacking into already-hacked infrastructures of social reproduction (a jugaad water supply) has been a common feature of everyday life well before the advent of the mobile phone.

In what follows, I draw from in-depth interviews with women and men about work, kinship relations, and mobile phone practices in Bhopal, Bangaluru, Delhi, and Mumbai. The interviews were conducted by myself and, more recently, by three research assistants, Shiva Thorat, Rachna Kumar, and Anisha Saigal, from 2010 to 2015. After a snowball sampling technique, we focused on questions of how differently positioned women and men were renegotiating older domestic and urban metabolisms through a relatively new technoperceptual assemblage: the human-mobile phone-communication network assemblage (Bittman et al. 2009; Castells 2006; de Souza e Silva 2006; Rangaswamy and Nair 2010; Strumińska-Kutra 2016).

In an interview with Anju, a twenty-year-old woman working as a domestic servant in South Delhi, questions of social reproduction and mobile technologies were tied together through the importance of maintaining family cohesion in the wake of her father's death.[3] The precarious economic situation of Dalit communities structures the specific challenges for unmarried women at such times of family hardship. In lower-caste, urban communities, the need to escape poverty by seeking out alternative economic opportunities often places women in danger of trafficking. However, as Meena Gopal has noted, the situation of poverty is complicated by numerous other gendered discriminations, such as child marriage, sexual exploitation within families, domestic violence, desertion and abandonment by husbands, denial of educational opportunities, and customary family practices, among others (2013, 239; see also Sen and Nair 2004; Vindhya and Dev 2011).

Anju related, "I've studied till seventh standard. My father passed away, as a result of which I had to drop out of school and get myself employment in order to help my family sustain. I picked up a job to run the house." It is important to note that for Anju the death of her father highlighted the life-changing inaccessibility to capital and resources in order for her to continue her education, and she had to seek employment in the social reproduction of middle-class, upper-caste households. This poorly paid and extremely grueling work is characterized by the daily callousness of the Indian middle classes toward their "servants," which outdoes the worst excesses of feudalism, according to the feminist legal critic Nivedita Menon. "The polite term 'domestic help' that has replaced the word 'servant' in public usage is perniciously misleading. Make no mistake: these are servants. They are treated as less than human, less than pet animals. Apart from facing physical and sexual abuse—which is common—domestic workers perform heavy unrelenting toil, for they have no specific work hours if live-in; no days off or yearly vacations if part-time. Not to mention the routine humiliation that is their due" (2011, 18).

Not surprisingly, the mobile phone ecology folds into the work of social reproduction, as well as posing a source of both danger and trouble for domestic workers and their employers. In the context of contemporary gender and sexual politics in India, the mobile phone is often seen as a threat to caste control and sexual contamination. As Menon notes of one case in Punjab, "Husbands and their family members suspect that the women are talking to other men over the phone. Even if the women are only calling their parents, said the chairperson, that's a problem, as being in constant touch with their natal homes hampers their adjustment into their new homes" (2011, 32; Menon and Nigam 2007).[4] In the context of domestic labor, the mobile phone allows Anju to be at least potentially in constant communication with her family and different employers. The timings of her work shift can be in relation to the convenience and emergencies of her employers: her employers, and their children, will call in advance and ask her to buy household provisions or run errands prior to her actual arrival; she can communicate with them about an absence or lateness due to illness or traffic (chronically bad in South Delhi) or any other logistical coordination. Second, the mobile phone allows Anju to have continuous access to media, which in turn helps to make her social reproduction work less tedious. As I had explored in *Untimely Bollywood* (2009), the practice of the "time-pass" (i.e., killing time) is an important aspect of neoliberal life in India; in Anju's ecology the mobile phone enables her to pass time during slow work periods. As the owner of a Vodafone feature phone, Anju engages in its basic functions "such as making calls, listening to music, taking pictures, recording media." This is not a trivial feature of Anju's work in social reproduction: the pleasures of her phone's media quite literally work around the exploitation, or "primitive accumulation," of the middle-class domestic space. At the time of the interview, Anju didn't know the process for accessing the internet or using Facebook, which means that hers is more a subaltern workaround ecology than anything narrowly digital. Also, there is a fairly seamless interface between old and new media in Anju's ecology: "I listen to music. I usually watch TV for that but in case the phone's around, I listen to it on my phone. The music is on the storage device. I access that" (although later in the interview she says that she "feels" there is a difference between TV and mobile). This (usually popular Bollywood) music comes about through paying for a specific piracy hack or jugaad: "For rupees 100, you can get 10–20 songs of your choice."

In Anju's practices, social reproduction and mobile technology link her work to her and her employers' domestic life through specific and differently scaled technoperceptual assemblages. These assemblages are also spaces of controlled autonomy and urban "metabolic imbroglios" (Sage et al. 2015; Swyngedouw 2006). In what sense could a city be thought in terms of metabolisms beyond the limits of the biological metaphor? Recently, the "economic developmental" rendering of urban metabolisms has become a context for the neoliberal capture of creativity. Richard Florida, in his "apology" for the creative clustering thesis, writes,

> Superstar cities have unique kinds of economies that are based on the most innovative and highest value-added industries, particularly finance, media, entertainment, and technology. Everything in these cities happens fast—information travels at lightning speed, innovation occurs at a rapid pace, businesses form and scale up more quickly—and this speed, along with their sheer size, underpins their advantage in productivity. That rapid pace is not just an impression one gets, as in the old cliché "in a New York minute"; it is an objective, scientific phenomenon. Scientists at the Santa Fe Institute, a think tank specializing in complex adaptive systems, have discovered that cities have unique kinds of metabolisms. In contrast to all biological organisms, whose metabolic rates slow down as they get larger in size, the metabolisms of cities get faster as they grow larger. With each doubling of population, the Santa Fe scientists concluded, a city's residents become, on average, 15 percent more innovative, 15 percent more productive, and 15 percent wealthier. (2017, 25, 46–48)

Here the supposedly objective authority of science is mobilized in an ideological obfuscation of violent processes of expropriation, segregation, and dispossession. Speed is confused with intensity as such, each superseded scale is immobilized, and causality is reversible: population growth spurs innovation, and innovation feeds a population spurt. As we have seen in previous chapters, the processes of value cannot be reduced to capitalist measure and strategies of accumulation; this is the revolutionary, monstrous promise of capital itself, the social and political creativity of a neoliberalized multitude-commoning (Hardt and Negri 2009).

In such informalized ecologies, horizontal connectivity and vertical scaling are enmeshed in flows of power, information, matter, and energy (Sage et al. 2015, 496). Anju remarked that "everyone in the house shares the same mobile phone. . . . We'd heard that this brand was good. We decided mutually amongst the members of the house and my brother had recommended this." "Hearing that the brand was good" suggests that already before the entrance of the handset into the house, the kinship resonates through experiences of branding and what marketers call electronic word of mouth (see Wang et al. 2012). The scales of these fluxes of neoliberal consumption have overlapping logistics, enabling the smooth movement of matter, information, and capital. Each overlapping scale expresses its specific power (essence) in different attributes and affordances, in practices, durations, codes, and networks. Everywhere, it seems, these powers serve to accumulate the arsenals of a militarized capital, everywhere deterritorialized but only partially autonomous commons work around and exit the biopolitical (Hindutva) security state, India's very own biostate. These commons are affectively mutating (and hence open to disruptive or even at times revolutionary practices of commoning) in and through digitally networked mobile devices. We return to the question of scale below, but for the moment I want to diagram further the dynamic space of subaltern political ecologies. Patterned but unpredictable, this terrain of neoliberal value cycles requires a genealogy of what Lefebvre called the construction of capitalist representational space—bedroom, dwelling, house, or square, church, graveyard—which is "alive: it speaks"; such space-times are the loci of passion, of action and lived situations, and may be qualified in various ways: "it may be directional, situational or relational, because it is essentially qualitative, fluid, and dynamic" (Lefebvre 1991, 42). Representational space is not a representation of space; rather, it is an affective, collective, and expressive space, and this living kernel of space-time is dynamic and qualified through sensory-motor circuits of repetitive, habituated work (cooking, calling, cleaning, washing), with its continuous durations of joy and tedium, reward and frustration. Again, these processes both blur the borders between home and the world and blur work and pleasure as both are differently repeated through this affective space, mobilized in different "smart" strategies that capture value, rationalize sense, and organize forces. The mobile phone is one constantly reinvented element, or flux-machine, of a more general tendency in an emergent political ecology in postcolonial India. These deterritorialized fluxes and displaced and decoded limits generate patriarchal anxieties of and sometimes violence toward the permeable borders of domestic space and those (women, children, elders) who would pass through them (Deleuze and Guattari 1983, 232; see also Ajana 2013; Lyotard 2014).

Sharing one handset among all the female members of the family is quite common in India, especially in subaltern urban communities. This also means that all media is shared by the household's women on one mobile phone, and the knowledge of its use is shared among them (Kubitschko 2015). "The media is common for us. . . . My elder and younger sisters taught me to use the mobile phone. I didn't know how to operate it; they learnt it amongst themselves by using it and later taught me." In this sense, Anju's ecology, her nonhuman agency and the consciousness of its powers, is distributed between her changing, emergent capacities and the designed affordances of the phone (and associated networks), and between herself and her siblings. It is this practice of sharing and commoning of digital media and the ethical know-how of its ecologies that points toward new practices that exceed the capitalist capture of value (Arboleda 2015; Berlant 2016; Hardt and Negri 2004, 2009; Varela 1999). The notion of commoning as a workaround through the mobile and thus a makeshift pedagogy between the sisters for its requisite hacking skills are effective and uncelebrated everyday practices for Anju and her family. These practices are ontological dimensions of her social reproduction. As she put it, "As long as [the jugaad, or hack] comes to be of use to others who are in need, it is good, regardless of whether this is unscrupulous or not. Otherwise, these hacks are useless." While this seeming virality of jugaad practices proliferating in informal ecologies of the multitude-becoming-commons presents a clear and certain image of practice for revolutionary becomings, my research has led me to question the postcolonial ontology of this image. Anju's ecology, her distributed agency, expresses itself in residual and emergent practices that join technology to perception through actual and virtual senses, values, and forces. This image of practice is itself a common notion—common, that is, to two or more multiplicities—of subaltern ecologies in specific ontological struggles.

Hacking Ghar/Bahir

Rekha lives in one of the many slum communities in South Delhi; she is eighteen. Her relationship to the gendering of her ecology performs and produces the exclusions of patriarchy in a way that situates her subjectivity in a hybrid, liminal contact zone of piracy, outsiders/women, desire, and (circumscribed) experimentation. In her home, "My brothers own a mobile each, personally. My sister and I use my sister in-law's mobile phone if we need to make calls, or anything else. All the men in the house have their personal cell phones. The ladies share a common phone." This gendering of mobile technologies follows, as we have seen, what Chatterjee has termed the "ghar/bahir solution" to the anticolonial women's question, and its contemporary digital blurring. Chatterjee noted that in the British colonial discourse that sought to ban sati (the sacrificial murder of a widow on a husband's funeral pyre), "effervescent sympathy for the oppressed" is combined with a moral condemnation of a tradition that was seen to produce and sanctify the murder of colonized women (Chatterjee 1989, 623). Thus, the nationalist resolution to colonial sympathy was to create a binary between the outside (the terrain of national struggle, of material progress and history) and the home (the realm of Indian identity and spirituality, embodied in women) (624).

The argument of this chapter is that contemporary subaltern ecologies in India have not only deconstructed this binary, but also have become the site of differently scaled experiments into what Gibson-Graham have called ontologies of the possible (2006, 2008). These care economies of social reproduction, the "arcana" of their ontological processes and solidary practices, have no fixed or digital archive (Deleuze 1992a; Derrida 1998). Its search/reconstitution will never break with representational metaphysics. Immense and immeasurable, the processes and practices constituting Rekha's social reproduction, the performance of knowledge, capacities, and media consumption repeats each time with a difference the sense, value, and force of the (post)colonial ghar/bahir. Rekha watches "whatever is coming on the television." These media streams often intensify and deconstruct the gendered binary in their narrative and perceptual form and ideological content (Mankekar 2015). These media effects are differently involved in the affective atmospheres through which Rekha affects and is affected by her ecology. She likes listening to music on her phone the best. Her process of acquiring music follows a habituated pattern of assembly and encounter (Arboleda 2015; Lefebvre 2003; Merrifield 2013): "First, I check with my friends, and ask whether he/she has the song that I want. If they do, I ask them to transfer it [through Bluetooth]. I enquire from almost everyone I know. If it remains unavailable, then I have to go to the local shop to get the song." Note that this initial step is informed by and produces her joyful passions: digital music, networked sociality, illegal downloads. In what way does Rekha's ecology transform these joyful passions into ressentiments separating her from what she can do or into common notions that actively experiment with (i.e., diagram) the becoming commons of late capitalism? Importantly, she later pointed out that her brother "purchases a memory card or a device; he purchases the music himself. Unless it's some extremely nice (popular) track, we usually never tell him what to get." Yet, "I do feel strongly that we should also learn it. I want to learn the process of getting the music but my brothers do not permit us." What are the contours of this gendered struggle around piracy know-how and hacking practices?[5] The media on the men's phone and the women's phone are different; their access to the skills required to hack the assemblage is explicitly controlled by male power in the family. Even so, Rekha has attended carefully to how her brothers acquire her media:

Q So say, you do find the track you're looking for, how do you then access it? Through Bluetooth?

A Yes, through Bluetooth.

Q Do you transfer the track on your phone?

A No, I have to ask my brothers. But I'm aware of how it is done.

Q So you know what goes in?

A Yes, you need to turn on the Bluetooth in both the devices and pair the phone. Once the phone is paired, the file can be transferred. I am aware of the process but I don't know how to do it. . . . Nobody taught me. It used to be kept around the house. While growing up, I used to fiddle in order to play games and switch between that and playing music. I learnt it on my own like that. I only use the phone when I want to listen to music. Otherwise, I don't touch the phone. I'm scared to use it beyond that because you never know when the phone might start malfunctioning if one continues to fiddle.

In Rekha's distributed agency, technoperceptual capacities coevolve in assemblages, which are the minimal units of agency. Her changing relationship to the mobile phone charts the struggles around neoliberal habituation and postcolonial social reproduction. This difference is in the affective transitions and dynamic spaces involved in playing or experimenting with the assemblage itself, "with everyone I know." It presents itself as a danger, but also an open, commoning secret. With different rhythms, Rekha's complex relations with her ecology perform the social reproduction of caste and masculine power and female incapacity, but in a way that allows technical discourses (pairing through Bluetooth) and commoning practices to become to an extent familiar to Rekha. This familiarity extends her practice within and against patriarchal power and the knowledge that ramifies it (Coleman and Golub 2008):

A Only my brothers know the process.

Q But these hacks can be observed and then turn better/improved upon. Do you think this should be shared with others?

A Definitely, but people (including my brothers) don't. They'll discuss it amongst the male members of the family but not with the outsiders or the women of the house.

Q Do you think jugaad can backfire/cause damage to those who do this? How?

A I do think it can backfire. A lot of people who hang around while trying to figure out a hack/jugaad can be manipulative. I think their [her brothers'] sense of direction can go wrong when others can go to any limit to get the result. I'm concerned about the safety.

Negotiating gendered power within the space of the home, Rekha develops an ethics of know-how (Bhabha 1994; Varela 1999). Rekha is aware of the risks, limits, and passages in its fluxes, and implicit in this knowledge are the ontological relations that such everyday hacks enfold in a necessarily distributed agency.

Commoning Jugaad

Hacks can common the resources of the multitude. They can create, in the words of Stefano Harney and Fred Moten, undercommons through fugitive study (2013, 28; see also De Angelis 2003, 2017; Lamarca 2014). Jugaad is a type of fugitive study. The tinkering know-how that is the processual life of all hacks is also an encounter with dangerous metabolic imbroglios, as we have seen, so its practice and communication must be careful. Visibility, as Foucault (1995) once warned, can be a trap. Turning to an example from the era of cable TV in India, in another interview (conducted by the Delhi-based researcher Anisha Saigal), Kanchan, a lower middle-class woman from Delhi, recounts her history of negotiating gender power through a safety pin and cable TV.[6]

> Q Have you ever used a jugaad besides your mobile phone, maybe in your family or somewhere else?
>
> A ... when cable TV was introduced in India I was a child. My father was completely against entertainment options like these, so I got around to using a safety pin to connect the neighbor's cable wire with our television set. I'd bring the cable wiring closer to our house to do that and we'd get clear transmission of cable in our house. I've done this personally and this is my own idea. This was an extremely successful idea. We've seen several films like this together on the cable through this jugaad. My neighbors had a cable television connection and my father had trouble accepting this in our house. The wiring for the neighbor's television would go via our window. Whenever my father would go to work, I'd take a safety pin out, pull the wire connection towards my television set and we'd get great picture quality and sound. . . .
>
> Q Did anyone teach you this technique/trick?
>
> A I tried this myself. As a child when I'd studied physics, I used to understand quite a bit and directly applied it to life. If I had to use a pin—as a metal in a wire . . . so things like that.
>
> Q Have you ever taught this or any other jugaad to anyone in your family or friends circle?
>
> A Yeah, I taught this jugaad to my little brother. Now the times have changed, there is dish television, etc. We don't need this jugaad anymore. (Kanchan interview, July 10, 2015)

In this example, Kanchan, as a lower middle-class urban adolescent girl, works around patriarchal controls of social reproduction by refusing to accept the father's ban on cable TV. In her affective atmospheres, morality (is globalized media good or evil?) is displaced in favor of ethical know-how: tinker with it till it works (Varela 1999). The outside (ghar/bahir) is forbidden, but once let in, it shifts perception, reorganizes quotidian rhythms, refashions self-image, and pushes against prior boundaries of social and gendered relations (Mankekar 2015). And why does it have to be paid for? Kanchan gives her brother a lesson in pirate ethics by teaching him the potential subterfuge in a repurposed safety pin. She experiments in the assemblages at hand. Kanchan acknowledges that with the advent of the digital, domestic space-time actually and potentially mutates (stealing satellite TV or free Wi-Fi access, for instance). At this early moment in Indian liberalization, the emergent and dominant media rhythms of the digital would be negotiated through the probehead of satellite TV. The affectivity involved in this volatile ecology mobilizes specific capacities of bodily attention (how is the flow of radio frequencies through coaxial cables from transmitter to neighbor engineered, and how and where could it be redistributed without notice?), metallurgico-electro-chemical experimentation with matter's membranes (safety pin, cable wires, metal, plastic, perspiring skin), and vision, sound, and tactility (more or less clear image and sound, manual manipulation of cable, internal wires, pin, etc.). What are the embodied rhythms of this hack?

In any jugaad whatsoever, a distributed yet virtuosic agency becomes in ontological experiments with metabolic imbroglios, or bottlenecks in logistical networks. The capacity to affect the cable and TV connection and to be affected by a new, secret, intensive assemblage of media technology that to an extent, and for specific durations, opens controlled domestic space affecting other family members and neighbors. The time-space of the domestic becomes to an extent more plastic or variable, making the scale of her jugaad indeterminate. This virtual (purely potential) and actual capacity shifts and enfolds the boundaries of the home and the world and the reorientation of perception itself.

The jugaadu (the performer of the hack) fashions herself into an individualist and a virtuoso ("I did it myself"), yet the material, intensive, and affective processes mobilized are preindividual and collective in that their relations form tendencies, affordances, effects, vectors, and events exceeding and virtualizing any given hack, while through these very modes always remaining susceptible to control. The new assemblage materially and epistemologically exceeds but is covered over by the hack in the sense that it is the labor of software engineers, cable TV installers, electronic components supply-chain workers, miners, and processing-plant workers that is covered over in the fetish of the hack as an individualized act of defiance and virtuosity, as product covers over process (Deleuze 1994; Massumi 2002; Whitehead 1979). The question from the view of the method of affective ethnography shifts from How did she do that? to Which creative acts diagram and affirm processes of commoning over the fetish jugaad?

The assemblage virtualizes the hack as new capacities emerge from the interaction among hand, sweat, TV, domestic enclosure, cable, and electronic signal. Tendencies to affect and be affected are experienced and conjoined in unexpected ways through the displaced fetish of the TV image. At a certain indeterminate scale, a new metabolic imbroglio has come into existence.

Conclusion

This chapter has sought to relate jugaad practice to biopolitical analyses of subaltern media practices and the feminist critique of social reproduction in the context of postcolonial displacements in Indian urban political ecologies. Contemporary populist technocracies are parasitic on the unpaid and caste-based labor of women. Here, affective ethnography sharpens critical attention on urban metabolisms. Indeed, this demands, as I have attempted to show, a careful attention to actual and virtual diagrams of the distribution of agency across ecologies, bodies, intensities, durations, and matter. This diagrammatic method critically and affirmatively engaged with intensive philosophy, cyberfeminism, and autonomist Marxism in the context of postcolonial India. The new lifeworlds coming into being today have to be situated, as these chapters have suggested, not in the mise-en-abyme of the aporia of life and nonlife, but of better and worse actions on the body's capacities to affect and be affected. This is ethics.

In India's neoliberal and entrepreneurial smart cities, mobile-based piracy processes help to shift the contours of what constitutes the ontology of urban political ecologies. Attending to the laboring conditions of differently positioned women within technologically transformed, and hence ontologically open, and varying domestic spaces affirms a potential politics of solidarity linking the commoning of digital ecologies to the overthrow of the primitive accumulation of patriarchal forms of capitalist social reproduction. In our examples, different metabolic imbroglios were created or experimented within the affective, ontological transitions necessary for the development of historically situated jugaad ecologies. Following Marx, Arboleda (2015) notes how general social knowledge (that is, "the general intellect") has over the past two centuries and globally become a "direct force of production" (Marx 1973 [1939], 706; see also Arboleda 2015, 37; Hardt and Negri 2009). This process has only accelerated into the twenty-first century, as the assembly line and domestic, agrarian, and wage labor have been increasingly refashioned by science and information in a constant search for new sources of wealth creation (Arboleda 2015, 37; see also Berardi 2009b; Hardt 1999; Lazzarato 2006). As we have seen in the context of jugaad ecologies of social reproduction in India, the passage to an information economy has "not only transformed production itself but also reconfigured the very nature of labour and the labouring process, as information, communication, knowledge, and affects are now foundational in processes of production" (Arboleda 2015, 38). The diagram of this ecology's becoming denaturalizes capitalist value as well as the (post)colonial and patriarchal binary of the masculinist world and feminized home. As practices of commoning resources of knowledge, matter, and care in and through mobile phone ecologies, they develop new relations of ecological "spatio-temporal rhythms of nature as transformed by a social practice" (Lefebvre 1991, 117). Drawing on the work of contemporary Marxist feminist and queer ecologists that take the ontology of social reproduction as their point of departure, I sought to bring the question of mobile phone practice into productive relation with matters of urban metabolisms of technology, kinship, care, and habits. Informed by postcolonial subaltern studies, this feminist research has put forth biopolitical and gendered genealogies of the contemporary regime of primitive accumulation in the domestic mode of production. The examples drawn from interviews in Delhi develop critical affirmations of workaround practices that sabotage, hack, and repurpose private property and patriarchal kinship, thereby creating conditions of encounter and experimentation with metabolic imbroglios in the interests of an emerging commoning of infrastructure and resources (Berlant 2016; De Angelis 2003; Lamarca 2014).

DIAGRAMMING AFFECT

Smart Cities and Plasticity

in India's Informal Economy

As we have seen, the aim of smart city design is to fold affective processes into globalized value chains through the intimacy of mobile devices. These value chains are themselves informal and formal ecologies of subsistence, production, and "smart" innovation (Arboleda 2015; Beer 2015; Birkinshaw 2016; Castells 2015; Clough 2010; Massumi 2015a, 2015b; Odendaal 2006; Thrift 2006, 2007). This work on smart urban design suggests that political ecologies of innovation and learning render time and space plastic, or ontologically variable, but that this variability is produced within the social synthesis of what Jasbir Puar has called debility (2017; see also Easterling 2014, 31; Dale and Latham 2015). Puar's concepts of debility and debilitization understand the particular processes of value extraction and population control under neoliberal capital as the very condition of possibility of affective ecologies. We have also seen that, given the feedbacks in these ecological processes, the debilitating "smartness" does not flow from policy to people, but habituates a recalcitrant multitude for capture (Culp 2016; Hardt and Negri 2009). This analysis pursues the implications of this field of relations for political ecologies of networked sociality. The affective geographies of embodied social rhythms and digital media assemblages have brought out the biopolitical dimensions of contemporary strategies of neoliberalizing the postcolonial city that focus on creativity and its modulation, disadvantage or precarity, gentrification and dispossession, privatization, "primitive" capital accumulation, and technologies of risk (Fraser 2008; Marx 1976 [1867]; Patel 2017). As we have seen, in India one of these neoliberal strategies of control and accumulation is the mobile phone–enhanced smart city, partly following its emergence in North American and European contexts of ecological sustainability, creative clusters, and info tech–based innovations in accountability, transparency, education, and civic participation (Florida 2017; Saha and Sen 2016). This chapter draws out the postcolonial urban processes, relations, and affects in neural and spatial plasticity in acts of learning, political practice, and contemporary smart city design (Braun 2014; Castells 2015; Changeux 2004; Connolly 2002; Deleuze 1992b; Guattari 2005; Parisi 2013; Savat 2009; Virno 2003; Thrift 2014; Grekousis et al. 2013). Looking at the urban contexts of Mumbai and Delhi, I show that widespread and striated workaround cultures in both cities, linked to the affective labor of women in social reproduction and informal subsistence economies in different ways, expand the scope and modulate the tendencies for urban and neural plasticity (Barbagallo 2015; Barbagallo and Federici 2012; Berndt and Boeckler 2011; Federici 2003; Sanyal and Bhattacharyya 2011). Practices in social-reproduction and mobile phone assemblages, media habituation, digital production and consumption, and social networked communication rearrange ecologies among people, things, perception, and procedures and require continual recalibration between the technologies of postcolonial governmentality and the populations modulated to produce specific orderings of the world (see Foucault 2003, 2007; Thrift 2014; Arora and Arthreye 2002; Arora and Rangaswamy 2013; Valayden 2016).

Building on feminist and postcolonial research on work and social reproduction in the so-called informal sector (around 90 percent of the national economy in India), I look at how hybrid practices such as jugaad suggest a prior field of subaltern recalcitrance and invention in practices of the contemporary informal sector (Birtchnell 2011; Hardt and Negri 1994; Kaur 2016; Lazzarato 2006; Radjou et al. 2012; Rangaswamy and Sambasivan 2011; Nagendra 2016). These practices come out of the ontological conditions of extreme resource limitations (the ongoing debilitization of biopolitical populations) in India, which differentially affect differently abled women, children, men, the elderly, and Dalit and minorities. A jugaad ethics is a becoming ontological of being itself. Antonio Negri, writing during his imprisonment, limns this ethics of composition as a question of liberation:

> The ethical act will thus be an act of composition, of construction—from the heart of being, in the tension between the singular and the collective. The possibility of a total violation of the world does not lead us to qualify action indifferently. The negation of every form of dualism and every mediation does not suppress the ethical alternative: it displaces it, resituates it on the extreme limit of being, where the alternative is between living and being destroyed. The radicality of the alternative highlights its drama, its intensity and irreversibility. And it is precisely, and justifiably, in this intensity and drama of the choice that ethics becomes political: the productive imagination of a world that is opposed to the world of death. . . . The productive imagination is an ethical power. (Negri 2004, 5)

These practices are both the targets and instruments in neoliberal strategies of measure, control, dispossession, depoliticization, financialization, accumulation, and extraction (Kaur 2016; Puar 2017; Radjou et al. 2012;). Here my itinerary follows through on a critical diagram of smart city discourses in India, posing questions of process, affect, potentiality, and habit on the terrain of everyday practice in the variegated economies and micropolitical formations of India's smart cities (Gabrys 2014; Latour 2005; Negri 1999, 161; Rossi 2015; Martin, 2013).

In what follows, I sift through the relational field of what I consider to be key dimensions of smart city agendas in India as they impact and develop feedbacks with informal urban economies and also produce experiences and ontologies of "metabolic plasticity." This chapter contributes to that literature in three main ways. First, drawing on research into subaltern social practices, and pursuing the implications for embodiment in the arguments of prior chapters, I show that workaround cultures in both media and social reproduction in India are prior to and continuously exceed forms of neoliberal habituation. Second, drawing on work in neurophilosophy and urban environmentality studies, I take questions of smart city design into a consideration of the ontological transformations in habituation and postcolonial urban rhythms. Third, linking this work on postcolonial ecologies of assemblages (agencements, or organizations of distributed agencies) to feminist political economy, I turn to questions of social reproduction and feminist organizing in the contexts of contemporary mobile media practices.

These ecologies have their own unevenly integrated flows of desire, value, matter, energy, media, information, and logistics (Massumi 2015b, 80–82; Rossi 2015; Williamson 2016). I turn first to a macrological contextualization of smart cities discourse as one of the many instruments of corporate neoliberalization and the elite globalization of Indian society that have evolved over the past thirty years. I then turn to a more focused consideration of the informal economy operating first in the media and cooking practices of a Dalit domestic worker in South Delhi, and, second, to the activism of the Why Loiter? project, which launched in Mumbai around 2009 and subsequently spread to other South Asian cities. I conclude with some thoughts on plasticity, creativity, value, and control.

Smart Cities, Creativity, and Metabolic Change

Consider, as a first approximation to the diagram of the mobile ecologies that I want to highlight, this excerpt from my fieldnotes from 2009 to 2010.

> Interview with Sanjay Bahadur (17, studied in Hindi medium till seventh standard) and Ranjit (19, studied in Hindi medium till eighth standard) at Dhoop Electronics. Both are from Dalit castes. I get on a motorcycle with the two of them and we drive off to Madanpur Khader Gaon (a fairly developed "informal" settlement in the Delhi-Noida area). In one of the major gulleys, in the middle of the block, there is Dhoop Electronics, open from 9 am to 11 pm. Sanjay and Ranjit were enthusiastic when they found out I wanted to know about mobile phone repair jugaads (workarounds). They sat me down, and started showing me how to repair a mobile: Integrated circuits (IC); dead plates or dead PCB (printed circuit board). . . . Basically they showed me the different parts of the PCB and what's on it: tiny little ICs. CUP IC, Power IC, Network IC, Hager IC, audio and keypad IC, flash IC (memory/software), display IC and camera IC. 3.7 volts is all that it takes to power up a mobile. Parts are found in Koral Baug, Gaffar Market, etc. What are the initial steps in starting up a dead phone? First, wash it in white petrol. Second, check SIM and connector; tracing, jumper wire, figure out if they can do the jugaad (their term). . . . There is a protocol in determining what the prob-

lem is, it starts with determining network connectivity. Check antenna switch; check all relevant ICs. A basic jugaad is "palti marna": switching components from one mobile to another; check the charging IC, usually has problems from stuffing the jack in too much . . . wipe clean the software giving on-and-off problems, hanging phone, file corruption (file urdjana, a "short-circuited" file) . . . the Vodafone dealer came by: Ranjit gave him a thousand rupees and he recharged for 1020; and all day people will recharge for 20 or 30 rupees, they make INR 20 on the deal. . . . Sanjay has been using mobiles for 3 years. He's had three permanent handsets in that time; has bought and sold around 7–8 phones; he likes the work it is interesting. For him, life with the mobile makes it easier to find and locate people; all work becomes easier, more suvidha (convenience). The phone has become necessary. Ranjit adds, "Everyone needs to stay connected; if you don't have one no big deal; but once you get one you can't live without it; for work: meetings, making money; time is very important, we have to budget our time." He uses it to stay connected to family and his girlfriend; work related calls are about 2–5 a day; and also to do schoolwork.

In this context, Ranjit and Sanjay use the practice of jugaad in a broader hacking ecology that includes the pirate hardware and software markets, the electronic junk–scavenging economies and the simultaneous neoliberalization and "informalization" of communication throughout Delhi. This hacking ecology traverses their domestic and work spaces. As two young, Dalit working-class men, their mobile phone ecology enfolds "service" labor, sexuality, kinship, and invention into a continuous process of (de)habituation. In these jugaadu practices, learning (and teaching) repairing mobile phones and living in its ever-extending ecology are about both making more money quicker and sharing digital communicative and material resources.

As we saw in previous chapters, mobile value-added services (MVAS) draw on these affective dispositions of jugaad ecologies in explicit ways, further entrenching mobile habituations in differently scaled smart city assemblages, even as market competition continually ratchets up the financial stakes. Consider this early excerpt from my fieldnotes with an MVAS executive:

In an interview with Rajiv Singh (September 29, 2009, in Delhi) I got a clear sense of how mobile telephony is changing India. First off, the mobile phone industry in India is the most competitive in the world, and this has had effects throughout the social body. Through tariff wars, giveaways, new handset development, MVAS, applications, games, and so forth, mobile phone operators such as Airtel, Vodafone, Tata-Docomo, and Reliance have had to find ever new strategies for perfecting their service deployment, increasing the reach of their networks, increasing their subscription base, and constantly lowering tariffs. Singh mentioned, for instance, Reliance Mobile's "Monsoon offer," where during the rainy season the operator had given away handsets for free. But since both tariff rates and call rates have been slashed, the average revenue per user (ARPU) has also been going down for some time, punctuated by doom and gloom forecasts from the industry pundits on a variety of media. Second, to maintain competitive advantage operators have resorted to developing innovative products and niche market segments. As for the first, it is widely noted that the Indian mobile market has been one of the most innovative globally in terms of what kinds of services and plans it offers to its subscribers: per second billing, sachet (i.e. low) pricing for prepaid recharges, lifetime free incoming calls for prepaid connection, one day internet connection plans, free night calling, free friends calling, etc. Such innovations have enabled the expansion of mobile usage among both new as well as mature mobile users.[1] With 95 percent of the mobile subscriber market in the prepaid category, the new, youthful user is a particular target of both MVAS companies and operators—Singh noted with a smile that young people seem to like Virgin brand phones. But also, much of the current efforts in monetizing the mobile market are geared toward the rural; the development of these niche markets has been greatly facilitated by new, cheap handsets which at the entry level run about "1000–1200 bucks." According to another MVAS executive, India is the only country in the world where you can get a camera phone for under a hundred dollars: the Fly brand (a European company) smartphone at less than Rs. 4000.

Fourth, and finally turning to MVAS, Singh said that "I would really call MVAS 'customer communication.'" For Singh, the mobile "ecosystem" (his word) is enabling greater and greater customer choice. Although the focus initially had been on viable tariff plans, MVAS is a rapidly expanding area of the mobile industry. The Internet and Mobile Association of India (IAMAI) noted that "The revenue through MVAS has been growing in the past years and by the estimates in the study, it is INR 11,680 Crores in size. Traditional MVAS like P2P SMS and Caller Ring Back Tones (CRBT) continue to be a substantial provider of these revenues. These basic services are expected to be large contributors. The stakeholders, however, are not able to garner higher revenues from value added services due to limitations in technological infrastructural platform. Further, in keeping up with the competition the call rates have been dropping resulting into low overall . . . ARPU. With the expectations of advantages that 3G could provide and the technological innovations (specifically in devices and content), the anticipations towards improved service delivery abound."[2]

Singh was more cautious in his estimation of the MVAS industry. Taking almost a fatalistic, and certainly an essentialist point of view, he said flatly, "Indians are not data friendly" (even at the time of the interview, this has not been borne out by shifting trends in increasing internet use or massive downloading of networked mobile games, although e-commerce growth continues to be slow in India for a variety of historical and infrastructural reasons). Singh felt that the best way for MVAS to increase its revenues and reach was by using interactive voice response applications through which companies have already seen a lot of uptake of services; also selling new tariff plans, vouchers, and access to utilities will also help. More, Singh was quite sanguine about something that will become more and more prominent in the years to come: mobile advertising. Through MVAS services, companies can engage in what is known in the industry as "below the line media," which is not full-blown advertising, but just little triggers of consumptions. For instance, after my daily health tip on my prepaid Airtel subscription, I get a message that reads: "Listen to live Cricket Commentary. Dial 543212607 (toll free). Charges Rs. 3/match day." Such below the line media, according to Singh, is bringing about a lot more usage, more "education or entertainment," through innovative strategies of communicating with customers. Turning his thoughts to the future, Singh notes that as the market expands and . . . ARPO . . . goes down, and as we see net revenue stagnating, the focus will shift to high-end big ARPO customers with 3G and 4G phones. Given that at present over 35 percent of mobile users cannot make a call due to their negative balance, there will be a gradual shedding of low-end customers (who in the industry are tellingly referred to as the long-tails, "bottom-of-the-pyramid types").

Between the subaltern distributed agencies of Ranjit and Sanjay and the global corporate strategies that Singh highlights in the early days of MVAS, biopolitical practices such as hacking and jugaad constitute the ontological conditions of ethics in a world of death and debilitization (Negri 2004; Puar 2017).

How can we pragmatically diagram these forces, senses, and values as they have coevolved with smart city policy in India? Jennifer Gabrys (2014) situates smart cities within a genealogy of neoliberal control and capitalist accumulation. Drawing on previous research, her work has shown how the ontology of urban processes and imbroglios are infused with and transformed by continually evolving computational, material, financial, and logistical processes in postcolonial smart cities (Arboleda 2015; Birkinshaw 2016; Galloway 2004; Hoelscher 2016; Rossi 2015; Saha and Sen 2016; Thrift 2014; Williamson 2016; Clark, N. 2017). As Gabrys notes, while informational or cybernetically planned cities have been under way since at least the 1960s, from the 1980s onward, this work designed "the plasticity of urban architecture" in zones of technologically spurred economic growth, in which cordoned-off digital cities remake "urban spaces as networked, distributed, and flexible sites for capital accumulation and urban experience" (2014, 30; see also Caragliu et al. 2011; Castells 2015; Easterling 2014; Glasmeier and Christopherson 2015; Graham and Marvin 2001; T. Mitchell 2008). Keller Easterling has herself pursued a diagrammatic method in her analysis of the political, economic, technological, and social tendencies of these special accumulation zones globally to form murky patterns of interaction (2014, 47; see also Bateson 1972). Nation-states "play a confidence game with the global market, announcing fluid plans and gambling heavily, perhaps even recklessly, on the zone or on new cities supported by zones" (Easterling 2014, 35). As mongrel forms of exception, zones slip in and out of visibility, and they shift between regimes of exploitation and financialization.

Lekki Free Zone—the largest free zone in West Africa and a double of Lagos—continues its expansion, with Chinese, not Nigerian, interests as the largest stakeholders. India is building the $90 billion Delhi Mumbai Industrial Corridor with nine "mega-industrial zones," high-speed rail, three ports, six airports, and a superhighway funded in large part by Japanese loans. On the other side of the world, Korean entrepreneurs have proposed a new science city called Yachay for the highlands of

Imbabura north of Quito in Ecuador. While the project has now been canceled, in April of 2012, Georgia announced plans for the new city of Lazika near the Black Sea. The customary promotional video scanned golf courses, fields of identical cartoon villas, and a cluster of skyscrapers that, sited on a swamp, would require foundations eighty feet deep. (Easterling 2014, 35)

Easterling's critical focus on forms of labor exploitation linked to exceptionalized zone-spaces brings out the solidary question of the smart city's modes of capturing value in these (in)formalized zones. Gabrys's interpretation of environmentality allows a renewed focus on the (at times failed) neoliberal capture of creativity in Indian smart city agendas and practices. The notion of creativity, or the creative but modulated event and unpredictable but probabilistic encounter, as one of the crucial attributes of neoliberal urban interaction, is legible in recent business management discourses on jugaad and frugal innovation, urban agglomeration, creative hubs, rentable work-spaces, smart regional corridors, and smart cities (GoI 2014b; Harvey 2005, 2012; Hepworth 2014; Hoelscher 2016; Lefebvre 1991, 2004; Parisi and Terranova 2000; Radjou et al. 2012). In India, one answer to managing the dynamic potential of urban creativity found in smart city documents is through arts and entertainment provisions. A different but related concept of creativity, one that smart city policy addresses implicitly, is at stake in the contextual neural plasticity inherent in the co-evolution of assemblages such as the human–mobile phone ecology (Clark 2003; de Souza e Silva 2006; Guattari 2005; Liang 2009; Malabou 2005, 2008; Massumi 2015a). I turn to the experience of creativity in these latter assemblages during the third part of this chapter.

As is well known and has been widely analyzed, India's experience of neoliberalism since 1991 has seen mixed results in terms of poverty and inequality reduction, labor conditions, jobless growth, populist democratization, and digital infrastructures and supply-chain resilience (Birkinshaw 2016; Carswell and De Neve 2014; Dey and Grappi 2015; Florida 2012; Gandy 2006; Lorey 2010; McCarthy 2005; Menon and Nigam 2007; Neveling 2014; Parikka 2012, 110; Prasad 2009). Labor migrations, regional disparities, deregulation, complex political formations, globalized assemblages of value extraction (from supply chains to risk assessments to financial derivatives), and rapid, sprawling urbanization have resulted in at times volatile but increasingly synergistic mixtures of transnational and indigenous capital, hybrid labor formations, and sharpened, often violent social and economic antagonisms. Following Sunanda Sen and Byasdeb Dasgupta (2009) and Kalyan Sanyal (2007), Ishita Dey and Giorgio Grappi analyze the facilitating role of the state, arguing that in India with the entry of private capital, "development in the postcolonial scenario" increasingly became "professionalized and bureaucratized; in other words, depoliticized" (2015, 153–154; see also Sanyal and Bhattacharyya 2009, 2011). Simultaneously, the Indian "entrepreneurial" state (in both its center-right and right-wing avatars) has overseen a more or less "silent revolution" in the information and communication technology (ICT) and telecoms sector for at least two decades, increasingly intertwined with the propagation of mobile-computing technology in the realm of urban policy, education, privatization, heritage, tourism, and commerce.

Smart city policy is one modality of that ICT revolution of habits in India. The ethical standpoint of the smart city mission is moored to the neoliberal legitimation of globalized capitalist commerce through projects of social and economic justice and environmental sustainability (corporate social responsibility).[3] As Kristian Hoelscher and others have elaborated in different contexts (2016, 29; see also Dey and Grappi 2015; Menon 2011; Menon and Nigam 2007; Sanyal 2007), Narendra Modi's socially conservative government embraces a vision of neoliberal globalization that projects a strong, business-friendly India outward, while managing increasing internal caste, class, gender, and sexual antagonisms through an unwieldy combination of censorship, subnationalism, financial/fiscal enclosure, patronage, populism, intimidation, and violence (Jaffrelot 2016; Khan and Ullah 2018; Menon and Nigam 2007; Phadke et al. 2011; Sanyal 2007).

In this context the neoliberal vector of depoliticization must be resituated to account for the complex relations of antagonisms of postcolonial Indian class, caste, and gender politics (Brown 2015; Dey and Grappi 2015; Mandarini 2005; Tronti 2005). The Modi government's Smart City Mission (SCM) aims to "create cities with smart physical, social institutional and economic infrastructure" (GoI 2014a; qtd. in Hoelscher 2016, 29). While there is little consensus globally or in India about the meaning of the term "smart cities," the entrepreneurial state has attempted to foreground various smart city instruments, including clean technology use as part of its Swachh Bharat (Clean India) initiative, differential diffusion of ICT, financing via public-private partnerships and private-sector investment, digitally networked citizen consultation, and "smart" or e-governance initiatives rolled out for urban local bodies (GoI 2014a; Hoelscher 2016). These smart city instruments assume an economic rationality of formalization, charged with integrating India's historically recalcitrant informal sector into new neoliberal circuits of measure and control. Hoelscher notes that "since its inception, the smart cities paradigm has been an elite-driven project focusing on private capital accumulation and urban, technology-led growth" (2016, 29; Easterling 2014). As an amorphous, context-dependent mix of institutional, physical, social, and economic infrastructures (GoI 2015), smart cities have been used to promote new and largely privately built and governed cities on greenfield sites, often in special economic zones (Hoelscher 2016, 29; Neveling 2014).

In the words of the Government of India, "Smart Cities are those cities which have smart (intelligent) physical, social, institutional and economic infrastructure while ensuring centrality of citizens in a sustainable environment. It is expected that such a Smart City will generate options for all residents to pursue their livelihoods and interests meaningfully and with joy" (GoI 2014a; GoUK 2016). How will India's smart cities measure residents' enforced joy (Culp 2016)? Mobile, digital, networked interactivity will be central to this measure and its indetermination. With David Beer, we see the emergence of Big Data and the so-called Internet of Things as the proliferation of "productive measures" responsible for "producing as well as tracking the social. They shape behaviours. As people are subject to these forms of measurement they will produce different responses and outcomes, knowing, as they often will, what is coming and the way that their performance will become visible" (Beer 2015, 10; see also Galloway 2004; Thrift 2014).

What, then, are the neoliberal tendencies of smart cities, understood as algorithmic control, and the intellectual property–driven informatization of everyday life? Three tendencies of class and caste stratification, namely private accumulation, dispossession and precaritization of the commons, and monopoly rent extraction, are decisive for smart city agendas. In India, as elsewhere (Caragliu et al. 2011; Slavova and Okwechime 2016), the pillars of a smart city are said to be specific institutional infrastructures that organize activities pertaining to governance, planning, and management of a city. Interactive mobile computing has, according to this state discourse, made this system "citizen-centric, efficient, accountable and transparent"; this infrastructure of interactive loops differentially connects, what James McCarthy, following feminist ecologists and post-

colonial critics, calls hybrid neoliberalisms across urban and rural, domestic and public space in India (Bhabha 1994; McCarthy 2005). More,
smart city physical infrastructure refers to its stock of cost-efficient and
intelligent physical infrastructure, such as the urban mobility system,
high-speed broadband infrastructure, the housing stock, the energy system, the water-supply system, the sewage system, sanitation facilities, the
solid waste-management system, the drainage system, and so forth, all of
"which are integrated through [the] use of technology." The smart city
extends to a definite management of the preindividual capacity for joy in
infrastructures of recombinant innovation of human and social capital
through education, healthcare, entertainment, and so on. It also includes
the smart design of performance and creative arts, sports, the open spaces,
and children's parks and gardens. Drawing on the deepening research into
the financialization of the Indian economy (see Münster and Strümpell
2014), we can today situate smart cities' economic infrastructures as development projects that generate employment opportunities and attract
investments, comprising "Incubation Centers, Skill Development Centers,
Industrial Parks and Export Processing Zones, IT/BT Parks, Trade centers,
Service Centers, Financial Centers, and Services, Logistics hubs, ware-
housing and freight terminals" (GoUK 2016, 9; see also Easterling 2014;
GoI 2014b, 3–6).

Official Indian sources "indicate that the growth rate of employment
in the organized sector has actually been declining, not only in the public sector but also in [the] private corporate sector" (Sen and Dasgupta
2009, 18; Birkinshaw 2016). Different forms of neoliberal measure imposed
on Indian workers must negotiate distinctly postcolonial hybrid articulations of subsistence, reproduction, pleasure, and work. Kalyan Sanyal
and Rajesh Bhattacharrya situate Indian labor formations in terms of the
"informalizing" tendencies of postcolonial capitalism in India, "in which
a large section of the population reproduces the material conditions of
their ever-precarious existence by engaging in concrete economic activities governed by a logic that is fundamentally different from the one that
animates the world of capitalist production" (2009, 36; Sanyal and Bhattacharrya 2011; Birkenholtz 2010). These informal ecologies are politically
ambivalent. On the one hand, by flouting regimes of copyright and resisting the neoliberal integration of labor through a proliferating and at times
nonproductivist informality, these ecologies moreover evolve practices of
potentially working around and resisting smart city agendas, especially

around intellectual property and land acquisition. On the other hand, by refunctioning and repurposing what are capitalist technologies—from mobile phones to databases—these ecologies habituate its practitioners and prosumers to neoliberal logics of Big Data and risk management (Amoore 2013; Culp 2016).

Plasticity in Jugaad Ecologies

I noted above that creativity, technical invention, and economic innovation are central to the neoliberal "table of values" (Brown 2015, 86) that govern the risk assessments of smart city discourses and practices in India. We can say, following Gabrys and others (2014; see also Foucault 2008; Massumi 2015a), that the smart city agenda aims to create an appropriate environmentality for developing capacities of creativity and innovation, throughout capitalist value chains, but especially in the creative clustered city itself (Florida 2012, 2017; Saha and Sen 2016).

The term "plasticity" is most actively used today in neuroscience, but it has found philosophical and geographical resonance in analyses of feedbacked processes of ecological or metabolic change in the city. In the rest of this section I make two overlapping arguments. First, neoliberalism tends to reduce neural plasticity to what Catherine Malabou calls the adaptability of patterns of habituation (2008; see also Lindstrom 2008; Zurawicki 2010). Second, neural plasticity relates directly to informal ecologies and their experimental ontologies of processes, tinkering, intensities, and becomings, in contexts of postcolonial neoliberalized precarity. These tinkering practices call for a renewed focus on habituation within informal ecologies that traverse smart city infrastructures across India (Bourdieu 1984; Butler 2006; Lévi-Strauss 1962; Lorey 2010).

In the smart city, the city is a "brain," increasingly available through material and virtual portals of experience accessed through mobile computing, connecting directly with the brains of citizen-consumers (Thrift 2014). For Michel Foucault, neoliberalism is a qualitatively new kind of conduct. The brain has become a site of struggle in generalizing this pedagogy of adaptability. Neuroscience (often of the most spurious sort) has become entwined with strategies designed to harness "human capital" through probabilistic models of the brain that coevolve with the government of human conduct (Brown 2015; Foucault 2008; Lindstrom 2008;

FIGURE 4.1. Smart city image, Mumbai (2016).

Prasad 2009; Virno 2003). In recent critical science and technology studies, neuroscience has been conceptualized in terms of its "numerous inter-
pretations, translations, and mediations" rather than an assumed "neuro-
realism" (Williams S. et al. 2011, 139, qtd. in Williamson 2016, 12). Here,
diagrams of affect relate the consumer-subject in the smart city to different gradients in algorithmic control, embodied feedback, and habituation
in experience economies. The social power of algorithms now plays a part
in organizing everyday life (Beer 2013; Deleuze 1992b; Thrift 2007, 2014),
and the algorithmic power of brain-based cognitive computing suffuses
social practices targeted for intervention through human–computer interaction with learning algorithms (Williamson 2016, 12). Indeed, a large
part of what it means to live today in smart city constructions across India
is to train oneself in pedagogies of canny negotiations of informal urban
plasticity. Through these (re)habituated negotiations, subjects constantly
update their own "conduct of conduct" toward the strengthening of core
competences and avoiding or mitigating the risks of encounter and assembly (Amoore 2013; Clough 2010; Clough and Halley 2007; Foucault 2008;
Grosz 2004; Nietzsche 1966).

Mapping Affects in Informal and Activist Ecologies

As we have seen, hacking ecologies have a common basis in the multitude's bottom-up biopolitical production—"languages, images, codes, habits, affects, and practices"—which "runs through the metropolitan territory and constitutes the very fabric of the modern city" (Hardt and Negri 2009, 250). The antagonistic contexts of postcolonial India and the politics of caste, class, religion, gender, sexuality, and environmental conditions shape access to and engagement with digital technologies of biopolitical production (e.g., the mobile phone). Across subaltern and neoliberal zones of contact and exception, forms of precarity, measure, and surplus value extraction govern more and more areas of social life in the targeted debilitation of biopolitical populations (Agamben 2005; Brown 2015; Grover 2009; Pratt 2009; Puar 2017; Rangaswamy and Sambasivan 2011; Sundaram 2009; Webb 2013; see also Heynen et al. 2006; Holifield 2009; Kaika 2005; Loftus 2012; Swyngedouw 2006). As such, smart technology is considered from the point of view of habit and plasticity in practices of controlled space-time. By and large, smart technologies in the city remain imperceptible to most people, except in the more or less conscious rehabituations required by their changing mobile phone ecology (Massumi 2015b, 75–76; see also Rangaswamy and Nair 2010; Venkatraman 2017).

Below, I focus on situating affect and agency in one interview with a Dalit domestic worker in Delhi, and a feminist hacktivist project of gendered spaces of loitering in relation to the earlier set of arguments.[4] In an interview conducted at an upper middle-class home in South Delhi, Leela, a sixty-year-old domestic worker and a Christian, originally from Nagpur but now a thirty-year resident of Delhi, and whose own children were raised far away by her extended family, recounts the day she was gifted her present phone. "It is a Nokia, 1260, something, I don't remember the number. I have a Vodafone SIM with a prepaid connection." Do you share the phone? "No, no, it is mine. I was gifted this by my nephew on Christmas Day. My nephew wanted me to have a new phone, and when he came to wish me on Christmas, he said, 'Moma, close your eyes,' and then he turned the mobile on and put it in my lap. There was a big noise, he gave me a big shock, and I asked him what he'd done: he said this is for you. I was so happy, it was an acknowledgment of the respect my family has for me." The affect of surprise, even shock, is doubled in the branded start-up jingle, in the act of respectful gifting, new intensities for the rituals of

Christmas. Leela used this phone for seven years, replacing the battery after six years, and she became pleasurably habituated to mobile phone games: "Nature Park. It's very enjoyable. Whenever I get a high score then I am 'very glad.' I feel joyous as soon as I hit a high score." Mostly, though, she uses the phone to stay in constant voice communication with her sister and extended family, and she uses the TV as her media consumption device. This longevity of ownership is unheard of in most of the urban mobile phone market today (two years with the same phone seems to be the national norm) but its anomalous nature is instructive. It pushes us to think the relations among embedded timescales of habituation (TV, phone, game board) and the material, logistical flows in the various assemblages expressed and actualized in Leela's ecology.

As the conversation turned to her lived conception (Massumi 2015b) of jugaad or life hacks, she, lowering her voice as if careful of what she was saying, begins to describe a practice through which one can say, "woh baat safal ho jaye" (may that problem be solved). But she insists that, for her, jugaad means no "ghalat kaam" (wrongdoing; we should keep in mind that she was being interviewed in her place of work). Have you ever taught someone jugaad? Extended pause:

> Yes. Unka jugaad successful ho gaya. Their jugaad was successful. God watches over us. My bank jugaad has helped with all the wedding prices, for the "shaadi, maine jugaad kar ke rakha tha, for the wedding I had prepared a solution in advance." My son always uses jugaads in his catering ka kaam, work. Everyone should do jugaad, because aaye ga kaam, hain na, aage mein? It will be useful [literally, work-to-come] in the future, won't it? If you sit down to work, a new jugaad will come in to your head [even as she mimics the employer's voice saying, "get it right"]. How can I retire and move in with my children? But these jugaads are lost when you go to sleep. Kitchen-wale jugaads, I don't share this knowledge, but I'll tell you a little something, pakodas are made soft by putting oil in the besan flour mixture.

Where do you learn from? "From anywhere. Sometimes through my phone, but often I sit by the person cooking, watch the timing of the ingredients. An educated person likes to learn new things. I like learning new things, it makes you more independent. . . . These 'kitchen' jugaads are not for sharing, I keep them to myself."

How can we think about the feedbacked relations between precarity and plasticity in hacking ecologies of the smart city here? This book has attempted to develop a typology of the two most decisive vectors: historically situated capacities and technological tendencies. Here, I want to suggest two overlapping affective terrains for further research. These terrains include the rapidly expanding capacity of communication and the increasing authoritarianism characteristic of the entrepreneurial Indian state's smart cities project in a time of war with the internal and external "Muslim threat," and the violent fragmentation and shoring up of Hindu chauvinist politics in a time of renewed Dalit militancy. Throughout India, subaltern groups have been empowered to share media and narratives that expose the strategies of marginalization, repression, and stigmatization that mark the Hindu nationalist evisceration of India's democratic and at least constitutionally secular civil society. Second, through and amid these biopolitical struggles, algorithmic control and forms of gendered labor habituate labor and consumption processes, while globalized value chains of social reproduction change relations within and between home and the world. The rhythms of domestic work in urban India are disciplined and modulated, intensified and potentialized in Leela's assemblage. Using different channels of communication open to her through her assemblage, Leela is able to manage the risks and boredom of domestic labor, integrating her habits into the informatized and informalized logistics of social reproduction in postcolonial Delhi.

I turn finally to the postcolonial feminist social movement called Why Loiter?, which was launched in 2009 after the publication of a surprising and innovative analysis into gender and urban space by associates of Partners for Urban Knowledge, Action, and Research (PUKAR, founded by Arjun Appadurai and Carol Breckenridge), Shilpa Phadke (an urban sociologist), Sameera Khan (a journalist), and Shilpa Ranade (an architect). In the subsequent seven years, the three published a book-length analysis of gender, caste, and religion in Mumbai; as well they actively contributed to the formation of feminist collectives "for loitering" in both India and Pakistan. From its inception, the Why Loiter? social movement activists have made innovative use of digitally networked media and devices in their different practices of hacking urban space (Jordan 2016). Why Loiter? blogs, tweets, group selfies, Facebook posts, and citizen journalism have highlighted the persistence of widespread rape culture in India, the stereotyping of women who loiter as sexually promiscuous, and the practical politics of solidarity between "tapori" (lower-caste and poor, masculinist) public culture and feminist organizing.

Why Loiter? shows how a strategy of reclaiming feminist and emancipatory solidarities in a time of violent Hindu nationalist and masculinist chauvinisms can disrupt the implicit securitization of smart city spaces for elite citizens and the explicit segregation of urban street life according to gender, caste, and religion. Why Loiter? is another way of posing the problem of experimental workarounds of the smart city's privatized enclosures, an insistence on a commons of encounter and assembly that affirms both a secular and spatial plasticity unique to urban metabolisms. As the authors put it in that early paper, socioeconomic transformations in Indian life after globalization have often "ossified hierarchical divisions" in the city not only to make it not just anti-all-marginal citizens but also, more important, to make their marginalization through social exclusion (for instance, in malls and multiplexes) and policing more acceptable; the administration of smart city policy through "slum demolition drives, the removal of street hawkers and the closure of dance bars" is based on prior marginalizations. The transformed figure of the respectable woman is as decisive in this biopolitics as it was during the anticolonial movement in the late nineteenth and early twentieth centuries: the most desirable among these women, the urban, young, middle-class, able-bodied, Hindu, upper-caste, heterosexual, married or marriageable women, become the established narratives of respectability in contemporary India. The bearer of all moral and cultural values that define the family/community/nation, a woman's "virtue, sexual choices and matrimonial alliances are informed by often strictly-enforced caste, community and class endogamy. In contrast, lower-class men are looked upon as an undesirable presence in public space. Their lack of legitimacy is underscored by locating them as a potential source of the threat faced by women, as putative perpetrators of sexual harassment" (Phadke et al. 2009, 187–188).

The practice that developed from this analysis, drawing on prior feminist organizing, initially centered on supporting networks of mostly middle-class, educated women loitering in public places, who posted photos and narratives on social network sites.[5] Over the subsequent seven years of affirming a postcolonial feminist "right to the city," the Why Loiter? movement has generated intense discussion within and beyond the movement by tying questions of security, risk, urban space, socially networked media

FIGURE 4.2. Why Loiter? image (2017).

practices, and practices of solidarity to a nuanced analysis of power and privilege and their antagonisms and intersections in contemporary urban India. The movement has inspired a play, several solidarity groups across India and Pakistan, thousands of Facebook and Twitter fans and supporters, and significant media coverage. The movement is a practical example of the use of mobile computing in life-hacking Mumbai's smart/Big Data city, and the development of practices tie local initiatives of feminist urban hacktivists to a broader social critique of gender, class, religious and caste exclusions in India.[6] Drawing on the work on biopolitical production from below (Ajana 2013; Arboleda 2015; Foucault 2007, 2008; Massumi 2015b, 77), I would situate Why Loiter? within a digitally networked relational field of the smart city that produces security through the "insecurity it's predicated on" (2015b, 77). The convivial work of feminist hacktivists, such as Why Loiter?, experiments with counter-ontopowers that break the "securitization loop" (2015b, 77; see also Arboleda 2015; Brown 2015; Ghertner 2014; Grosz 2013). Jugaad or hacking infrastructures, pirate and makeshift logistics, and commoning practices, most directly tied to the production and circulation of value in familiar forms—marriage, death, domestic labor, work, money, capital, power, time, property, and savings—and also in momentary space-times of experimentation with future-oriented capacities, both suggest politics for other futures and affirm the plasticity of the present.

Conclusion

This chapter considered the control technologies of India's smart city from the perspective of neural and urban plasticity in informal economies and activist practice. The control mechanisms of affect in smart city feedback in to the experience and practice of plasticity in urban space and habituation (Culp 2016; Varela 1999). Easterling diagrams these mechanisms through multiplier and switch technologies, which are "common active forms" in infrastructure space.

> An interchange in a highway network acts like a switch. A dam in a hydrological network, a terminal in a transit network, an earth station in a satellite network, or an internet service provider in a broadband network are all switches. Like the ball on the inclined plane, they establish potentials. Like a valve, they may suppress or redirect. The switch may generate effects some distance down the road or the line. It is a remote control of sorts—activating a distant site to affect a local condition or vice versa. Exceeding the reach of a single object form, the switch modulates a flow of activities. However deliberate the activities of the switch, it cannot control all of its own consequences any more than one could account for every use of the water flowing through a dam. Infrastructure space is filled with switches and remote controls, most of which are also multipliers repeated throughout the system, and tuning these active forms tunes the disposition of an organization. (2014, 44; on diagram as method, see pp. 47–49)

Smart cities, even if their modalities remain largely hidden as in most Indian cities, organize affective dispositions. The evolution of smart city agendas in India has been driven by forms of elite neoliberalism common to top-down smart city design the world over, but it also has proceeded through complex negotiations with organized and unorganized resistance to dispossession, caste- and religion-based chauvinisms, and neoliberal value extraction dominant in today's globalized regimes of capitalist accumulation. While the smart city is explicitly designed to tackle inequality and extreme poverty, its mandated enfolding into corporate copyright, monopoly, and licensing practices, and its biopolitical security already ensure that smart equals a proactive and entrepreneurial disposition. These create action-potentials within and against economic, informational, and military forces of the nation and the global North. Processes of capital flight, shifts in geopolitical alignments, gentrification by dispossession, and ongoing monopoly rents structure access to the smart city unequally. Anomalous habituations such as Leela's suggest lines of flight exiting contemporary smart technologies, and they intensify and reterritorialize the mechanisms of biopolitical control across all aspects of life. The convivial practice of Why Loiter? is a tactical refusal of this patriarchal and neoliberal control. Key here then is the potential of strategic bottlenecks, sabotage and repurposing, short circuits and embedded timescales in the informal jugaad ecologies to effect change in contemporary arrangements and practices of solidarity (Mandarini 2005; Negri 2004; Tronti 2005). In their own arguments, Andrew Culp (2016) and Kristian Easterling (2014) suggest that such hacks tend to be short-lived, dangerous, hypermanaged, and structured through different forms of dominance. At the level of policy, discourse, and practice, design targeting neural and urban plasticity captures potentialities emerging from the informal ecologies of biopolitical reproduction. The precarious logistics of these hacking ecologies take many forms but are ontologically one in that their material processes of energy, information, and affect form ecologies of practices operating across feedbacked and turbulent, connected, and controlled relations that are external to their terms.

I have suggested that in hacking ecologies visibility is a trap (Deleuze 1992b; Foucault 1995; Massumi 2002; Poster and Savat 2009). Leela acknowledges this when she lowers her voice as she talks about her everyday hacking, and a constant struggle around visibility is discernible in the various definitions of jugaad implicit in Leela's narrative. Conversations with mobile phone users such as Leela present several problems, explored in contemporary analyses of digitized environmentality and lived time-spaces. The first is the impossibility of the separation between ontopolitical environments and lived experience. Every event, whether of economic transaction, social reproduction, or urban hacking, is an embodied occasion for registration; every event proceeds either directly or is a workaround of digital registration. Leela is an acknowledged virtuosic cook whose experimentations in her own capacities is circumscribed by the precarious ecologies and durations of the Dalit working class, the durations of habituations within and against her own domination: she thankfully refers to her employers as her parents, the father as Papa, while practicing workarounds anticipating their disciplinary voices, which she repeats to herself. For her, the smart city traverses the rural and the urban everyday as conversations with her relatives, and hence reactivated memories of them, relate her to rhythms lived in villages and towns far from South Delhi. Second, new critical assessments of the effects of these digitized environmentalities and the forms of productive measure that are evolving today return us to the intersecting problems of control, extraction, and dispossession as well as the new forms of social media and urban activisms working around male, caste, and state control of urban space such as Why Loiter? (Savat 2009). These political ecological experiments pose the ontological conditions of domination and emancipation from the framework of the capacities of a counterpower to capitalist habituation in the smart city.

A Series of Minor Events

The Minor Events were never serialized but saw a kind of memorialization in an augmented reality show, where the show's admins followed the lives of these virtuosic geeks negotiating different technoperceptual mutations in themselves and others. The audiences organized into voting blocs of sensation, interactively affecting outcomes and dispositions, and so they "participated" in the experience of the mutations. It was a laugh, as they used to say. They turned it into a multiplatform game for augmenting what they wanted us to believe was reality.

But this media capture missed the diagrammatic potential of the part of the Events that exceeded their actualization (e.g., the immeasurable careers of virtual capacities to affect and be affected). This was behind the strategy in the jugaadus' insisting that we keep referring to those times as the Minor Events. Why minor? The Minor Events could never form a whole history, and yet there was a certain unity to them; they were not from the past and pointing to the future, but something that seemed to circulate by suspending time and all its values. The euphoria, delirium, even conviviality of hacking ecologies, the sheer creativity of jugaad practices developed a method of potentializing the always already plastic present, that passage from the lived present to its potential, that shock to thought and body, that flux, desire, and duration that arrive only and forever as the Untimely: something of that was lived in the Minor Events. What happened, first in Bhopal and then elsewhere, was the emergence of a practical path out of the postcapitalist morass of hypernationalisms, religious chauvinisms, and dogmatic populisms, not a shining path, not a New Way Forward, but something anomalous between the paving stones, something quite literally limning the actual formations of digital control in the virtual capacities experimented in cyborg becomings transvaluating all values. As the technological and social walls between Old Muslim and Administrative Bhopal began to crumble with the mutations in the techno-chemical phylum, an ecology in phase transition, when communities that had seemed to forget the intense solidarities of the secular nation for consanguinity and purified religion suddenly found themselves having to create together new terms of reference as common ideas, that is, ideas common to at least two multiplicities: the body and the mutation (well, three, then: postcapital). These capitalist and postcapitalist technologies of carbon- and silicon-based life had been hacked before, but rarely, and never with this kind of collective determination, in articulated commoning practices of global solidarity. That all felt very new.

What felt old were the identities tied to a supposed revolutionary or fascist past, to the once and future glory of a purified nation or religion, and the sense that the shift to an informatized life within an Anthropocene of volatile, far from equilibrium chemico-thermal ecologies had mutated sociotechnical life to the point where the established binary identities of race, gender, sexuality, and ability could themselves no longer be recognized or resembled. These mutations in sociotechnical life seemed quite small scale—isolated in neo-imperial Zones of Free Exclusion, such as Bhopal, the new capitol of the Hindu biostate, or in specific, quarantined populations exceptionalized in the shadow-margins of the overdeveloped centers of oligopoly. But there was a minor becoming that heralded an Untimely ecology, the jugaadu. The hacker penetrates, while the jugaadu flows like water, and always from the middle. It was this method of fluidifying infrastructures of capitalist command and control that became the abstract diagram, without resemblance or contiguity, of new becomings throughout the postcapitalist world. Jugaad practices necessitated a politics of plasticity for radical communities and political formations that increasingly dispensed with the identitarian performance of traumatized belonging marking all "minority difference" and developed an empiricist experimentation into the technoperceptual assemblages of globally distributed, but locally lived ecologies of commoning: race racing, queer queering.

BHOPAL REMAINED ITS OWN ecology of sensation, for me and for those I remember. The hills and lakes were still there. In the monsoons, the dark clouds seem so close in Char Imli Hills, where they put the camp for experimenting on the mutations, some of whom went on to reality TV stardom. In the beginning it was one of SourceStyle's sub-brands administering it, which was itself managed by a joint team of the Social Responsiveness Department of the municipality and a semi-private Belgian concern. Agents, and reagents, carbon dioxide and oxygen, heat, more heat, burning skin turning hyperconductive and radiating pulses of feedbacked information, testing the limits of volatility in an emergent chemical phylum: What can this new body do?

Conclusion

JUGAAD JUGAADING
Time, Language, Misogyny in Hacking Ecologies

Jugaad time is within, against, and outside historical time (it still doesn't add up!). Allied with the anomalous experience of the Untimely, jugaad time is a durational passage from one affective state to another that momentarily suspends the solidity of social and economic relations and of intellectual and manual labor, an interstitial time of material and affective flux (suspension ≠ statis), a time lived against the tables of measures, regimes of signs, and infrastructures of control of Indian neoliberalism. Antonio Negri reminds us that "Spinoza's philosophy excludes time-as-measure. It grasps the time of life. This is why Spinoza avoids the word 'time'—even when establishing its concept between life and imagination. Indeed, for Spinoza, time exists only as liberation. Liberated time becomes the productive imagination, rooted in ethics. Liberated time is neither becoming, nor dialectical, nor mediation, but rather being that constructs itself, dynamic constitution, realized imagination. Time is not measure but ethics" (Negri 2004, 5–6).

Jugaad time births forth a constructivist time of ethical know-how, that is, a time both constructed and emergent (Massumi 2011). At the same time, as a practice strongly associated with the hacking workarounds of lumpen, working-class, Dalit, and debilitated subjects, its media and business uptake is very much a historical phenomenon. The nonlinear history and pragmatic philosophy of jugaad give contemporary radical hacking practice one methodological resource immediately: the ethical know-how of the plasticity of political and technological ecologies (Guattari 2005; Varela 1999). I have considered the changing resonation among different types of plasticity, neural, affective, and urban. Drawing on some key lines of flight demarcated here, I want in these concluding remarks to specify the problems that jugaad practices help to pose better.

Jugaad, Language, and Specificity

At a Historical Materialism conference in Delhi, a year after the gang rape and murder of Nirbhaya in 2012 and a year before Narendra Modi's formal electoral victory for the BJP, when cultures of rape were disturbingly associated with jugaad practices (see below), in a collaborative paper Stefano Harney and I affirmed the micropolitical combinations of jugaad while also analyzing the necessary but as yet not sufficient critique of capitalist logistics. When I mentioned jugaad, the South Asians in the audience visibly changed their posture and expression, some leaning forward with keen interest, and others sneering with quick disdain: jugaad is a Punjabi/Hindi word with a specific affective charge. How is the question of language and affect sharpened in jugaad practice? Jugaad, a specific word in a hybrid language, is an ontological event, an encounter of creative plasticity in relations external to their terms. Gilles Deleuze, pursuing his ongoing critique of psychoanalysis, reminds us that what Félix Guattari understood by the question of language involved a diagrammatics (Deleuze and Guattari 1987, 105–111; Deleuze and Parnet 2007, 118) or a kind of pragmatics of language. Deleuze notes that for Guattari pragmatics is the decisive framework because it activates experiments in the micropolitics of language. Also, while there is no "competence" separate from "performances" in language, a pragmatics links these performances to their ecologies. Rather than abstract machines internal to language, such a politics diagrams abstract machines "which provide a language with a particular collective assemblage of enunciation (there is no 'subject' of enunciation), at the same time as they provide content with a particular machinic assemblage of desire (there is no signifier of desire)" (Deleuze and Parnet 2007, 116; see also Deleuze and Guattari 1987, 361).

Diagrammatics takes language to be a material force external to its terms, one that can be engaged in projects of deterritorialization, language refunctioned in a pragmatics of becoming minor, to become a foreigner in one's own language (Deleuze and Guattari 1987, 292). Each performance of the material and conceptual relations of jugaad is a specific intervention in a field of relations. Thus, its diagram understands the sensory-motor circuits of language and representations through their specific feedbacks with ecologies of sensation. Jugaad diagrams affects as capacities, affordances, and durational transitions that undergo the force of heterogeneous relations of power and their infrastructures. The value of jugaad, its immense experimentations and its immeasurable durational transitions, displaces and exit from capitalist valorization, show it to be parasitic on the autonomous and creative production of effective and collective difference—commons commoning (De Angelis 2007, 2010, 2017). The everyday experience and communication of the intolerable debilitation of biopower and geontopower, their distribution and application in intensive and material forces expands critical attention to energy grids, spit, sweat, rain, heat, traffic, and perception. All these forces have different trajectories, scales, affordances, temporalities, speeds, and durations: some are resonant, some are antagonistic, some resonate in an abyss of dialectical opposition, and some, at a given moment in time, are indifferent to each other. In frameworks explored in subaltern studies by Ranajit Guha and Gayendra Pandey and the postcolonial deconstructions of the subject-supposed-to-know of Gayatri Chakravorty Spivak, Homi Bhabha, M. Jacqui Alexander, Jenny Sharpe, Chandra Mohanty, Dipesh Chakrabarty, and others, this diagrammatic history of subaltern practices poses the question of language and more broadly of representation as such in terms of class, gender, and caste exploitation. Jugaad is a fractally styled answer to the "no" of both state (law) and capital (private property): work around it, hack past it, become indiscernible and revolutionary.

Jugaad and Revolutionary Becoming

How does jugaad contribute to the ongoing struggles for noncapitalist, anticapitalist economic equality and democratic social emancipation in India and indeed globally? The term "intersectionality" has come to constitute a mapping of power after neoliberalism. Jasbir Puar's critique of this formation is salutary (2017): the intersections can merely repackage the liberal essentialisms of the assemblages of the identikit. Her work makes clear that the conditions of jugaad are not one of hope (Deleuze 1992a). The conditions of jugaad's emergence—poverty, exploitation, debilitation, apartheid, malnutrition, stress, debt—suggest a field of subaltern recalcitrance, the diagram of which can potentialize futures. The entrepreneurial state that Modi's chauvinist hegemony has fostered in several key states has specifically targeted sites of radical solidarity politics, labor counterpower, and Dalit refusal. Throughout this study, I have avoided linking jugaad to one political project in India (e.g., Dalit-Marxist, queer, or feminist) or globally (e.g., autonomous hacking); its performance is and will remain an expression of a singular multiplicity of values, senses, and forces. That it has become a kind of international lingo of business innovation is one expression and discourse of this multiplicity. As we have seen in this book, jugaad as an event or encounter of creative plasticity in social and material ecologies can have several overlapping antagonistic and divergent senses and politics. If there is a politics of jugaad, it will be one that will be composed in coevolving and singular domains, and if there is a collective becoming implicated in this micropolitics it will have been a proleptic people commoning with each performance of jugaad, as in a revolutionary becoming.

I have for some time grappled with just what Deleuze meant by the term "revolutionary becoming." I return to it today in light of the slow learning that his work has initiated in me as a kind of intensifying propulsion (1992a; see also Bergson 2012, 366–369). The times are grim; at the time I wrote this, the worst mass murder in postwar American history was in all the news. Las Vegas didn't seem to have changed much two weeks later, but the timelines of such events, the scale of their effects, cannot be written in advance or in linear time. Yes, the times seem to herald the end of things as we know them. Perhaps we need a diagrammatics whose measures of better and worse are themselves gradients of intensification in thought and embodiment: a political diagrammatics, not necessarily more, or faster, but absolute acceleration itself—sensation, movement, resonance. I have followed the ontological implications of developing commoning diagrams in solidary assemblages: the diagram is in service of an untimely image of the jugaadus' war machine. There is good reason, rooted as always in a deadly alliance between the joyous and sad passions, to question why connectivity should be thought of as something authentically affirmative or truly critical for that matter. Have we presented yet another interpretation of the world, losing sight of what it means to transform it? Here politics must be understood to proceed like strategic cuts, or breaks, constantly at the intersections of processes of power, mobilization, and exploitation (Moten 2003). Andrew Culp's recent critical appraisal of the naive spontaneity of the connectivism of some Deleuzian studies should be read with care: the Untimeliness of a hatred of our capitalist debilitation cannot be overstated or stated in enough forms. Indeed, in many circles "revolutionary becoming" is the object of much ridicule today, reduced to a trite, ineffectual, and naif politics. Why has our time dismissed it so easily? Such passages are today quite familiar:

> Health as literature, as writing, consists in inventing a people who are missing. It is the task of the fabulating function to invent a people. We do not write with memories, unless it is to make them the origin and collective destination of a people to come still ensconced in betrayals and repudiations. American literature has an exceptional power to produce writers who can recount their own memories, but as those of a universal people composed of immigrants from all countries. Thomas Wolfe "inscribes all of America in writing insofar as it can be found in the experience of a single man." This is not exactly a people called upon to dominate the world. It is a minor people, eternally minor, taken up in a becoming-revolutionary. Perhaps it exists only in the atoms of the writer, a bastard people, inferior, dominated, always in becoming, always incomplete. Bastard no longer designates a familial state, but the process or drift of the races. I am a beast, a Negro of an inferior race for all eternity. This is the becoming of the writer. Kafka (for central Europe) and Melville (for America) present literature as the collective enunciation of a minor people, or of all minor peoples, who find their expression only in and through the writer. Though it always refers to singular agents, literature is a collective assemblage [agencement] of enunciation. Literature is delirium, but delirium is not a father-mother affair: there is no delirium that does not pass through peoples, races, and tribes, and that does not haunt universal history. All delirium is world-historical, a displacement of races and continents. Literature is delirium, and as such its destiny is played out between the two poles of delirium. Delirium is a disease, the disease par excellence, whenever

it erects a race it claims is pure and dominant. But it is the measure of health when it invokes this oppressed bastard race that ceaselessly stirs beneath dominations, resisting everything that crushes and imprisons, a race that is outlined in relief in literature as process. Here again, there is always the risk that a diseased state will interrupt the process of becoming; health and athleticism both confront the same ambiguity, the constant risk that a delirium of domination will be mixed with a bastard delirium, pushing a larval fascism, the disease against which it fights—even if this means diagnosing the fascism within itself and fighting against itself. (Deleuze 1998, 4–5)

For Deleuze, there are becomings that are silently at work, "which are almost imperceptible." Becomings cannot without violence be assimilated to linear history, whether personal or universal. Rather, becomings belong to "geography, they are orientations, directions, entries and exits. There is a woman-becoming which is not the same as women, their past and their future, and it is essential that women enter this becoming to get out of their past and their future, their history. There is a revolutionary-becoming which is not the same as the future of the revolution, and which does not necessarily happen through the militants" (2007, 2).

What does Deleuze mean by a revolutionary-becoming? Its method traces a minoritarian line, follows a channel, advances the imperceptible, but in a politics that as yet has no name, a monstrous politics of solidarity? In what way is this becoming revolutionary if it has no "future" or "past," merely passing between the two? More to the point here, in what way do jugaad diagrams extend a line of flight that both becomes revolutionary and affirms revolutionary becoming? I have avoided in this book, to the best of my ability, engaging in academic polemics on the limits and conditions of possibilities of the dialectic, or identity, or representation. I have tried as consistently as possible to develop a style of analysis that works by a kind of fractal becoming in conjoined sets of affective multiplicities.

But the embryo, evolution, are not good things. Becoming does not happen in that way. In becoming there is no past nor future—not even present, there is no history. In becoming it is, rather, a matter of involuting; it's neither regression nor progression. To become is to become more and more restrained, more and more simple, more and more deserted and for that very reason populated. This is what's difficult to explain: to what extent one should involute. It is obviously the opposite of evolution, but it is also the opposite of regression, returning to a childhood or to a primitive world. To involute is to have an increasingly simple, economical, restrained step. (Deleuze 2007, 29)

We have in this book been engaged in a kind of becoming-minor of ju-gaad, found in its collective interstices, in its commoning values something emerging between the paving stones of capital's formal paths and algorithms, a practice silently affirming an ethics of assemblages that experiments with a kind of commoning empiricism that invents "a people who are missing."

Jugaad and Precarity

Autonomist Marxists, queer scholars, and postcolonial critics have posed the relations between neoliberalism and the swelling ranks of the precariat in different parts of the world. Jugaad is a practice of the precarious, a practice of those for whom indebtedness and debilitization are forcefully correlated.[1] Jugaad, even in the celebratory hype of management consultants, helps to further demystify the relations and processes of domination, the dispossession and indebtedness at the heart of contemporary capitalist valorization. Jugaad, if understood as an expression of solidarity among the dispossessed, is a refusal of enforced austerity and the imposition of differential states of precarity all over the world. However, it is also an "innovative" working around those structures, and in that sense of value creation performs a legitimation of entrepreneurial value capture that continually restructures contemporary forms of affective labor and neoliberal management. Let us remember that the conviviality of the sharing economy is the disavowed modality of place-branding strategies in disparate creative industries discourse (Andres and Chapain 2013; Boltanski and Chiapello 2007; Pratt 2009).

What is the status both culturally and organizationally of the creative event in jugaad practice in the context of globalizing capital, nationalist chauvinisms, the normalization of rape culture, the everyday glorification of specifically urban expressions of masculinist heteronormativity, and caste-, region-, religion-, and language-based antagonisms throughout India? What can jugaad practice offer movements for feminist and Dalit social and cultural spaces in diverse economic and political contexts? Does precarity—as discourse, as concept—speak to the specificity of Indian jugaad practice? Does jugaad as a common notion, that is, a notion that is common to at least two multiplicities (or three: a diagram common to the body, its reproduction, and the Obstacle, for instance), travel beyond the supposed national confines of India? This is not a question of cultural appropriation but of radical strategies of developing solidarities in the ongoing struggle against all forms of oppression, and for the mutual emancipation of diverse, intercalated ecologies. Jugaad happens everywhere on earth. Everyone has at one point or another been a jugaadu, and you will be again. Postcapitalism will come to capture the revolutionary becoming in jugaad itself: that is without doubt. Is jugaad, as refusal of law, normality, hierarchy, and the status quo, simply a part of the shared heritage of those dispossessed by what was promised them? Clearly, as we have seen, one function of management discourse is to universalize innovation strategies—they should "sustainably scale-up." But no single jugaad scales up without a strategic infusion of private or public capital and resources; everyday jugaads become brands through financing, pitching, standardization, and monopolization (Airtel is an example). Indeed, an elaborate formalizing apparatus is created in and through the administration of jugaad cultures, for instance in smart city policy. Rather than romanticize the informal, the question from the perspective of radical ecologies of commoning and solidarity is what lines of flight from contemporary capital increase collective capacities for a noncapitalist landing? (see De Angelis 2007, 2017).

The ontological conditions of precarity, moreover, long predate the coming of neoliberalism to India; forms of feudal indenture (bonded labor) and capitalist indebtedness (moneylender debt) have long mixed with caste, class, and gender domination and exploitation in various parts of India, and they now exist side-by-side with the emergent relations of power in microcredit self-help groups throughout the neoliberalizing economy. In this study, I have considered the question of precarity from the point of view of the ongoing debilitization, refusals, and constructions of subaltern ecologies of hacking, bottlenecks, logistical imbroglios, social reproduction, and creativity in technoperceptual assemblages (Deleuze and Parnet 2007, 96, 123). We are constantly in these processes being separated from what we can do through a certain fetishism of joy, the enforced joy of postcolonial capital.

Funda kya hai? What's the Idea? Jugaad, Innovation, and Creativity

Recall Nikhil's light-bulb expression in the Sulekha ad. That expression is a joyous idea emerging in the midst of the background noise of habituated practice. This is also something that jugaad can offer to diverse milieus of radical hacking: virtuosic creativity. What does it mean to have an idea within jugaad practice?

> I could ask the question a different way. What does it mean to have an idea in cinema? If someone does or wants to do cinema what does it mean to have an idea? What happens when you say, "Hey, I have an idea?" Because, on the one hand, everyone knows that having an idea is a rare event, it is a kind of celebration, not very common. And then, on the other hand, having an idea is not something general. No one has an idea in general. An idea like the one who has the idea—is already dedicated to a particular field. Sometimes it is an idea in painting, or an idea in a novel, or an idea in philosophy or an idea in science. And obviously the same person won't have all those ideas. Ideas have to be treated like potentials already engaged in one mode of expression or another and inseparable from the mode of expression, such that I cannot say that I have an idea in general. Depending on the techniques I am familiar with, I can have an idea in a certain domain, an idea in cinema or an idea in philosophy. (Deleuze 2007, 312)

The dangerous thing about jugaad, of course, is that it is not an idea in any specific domain such as film or mathematics but rather a method, or an idea of the idea of working around (Deleuze 1992a). And so the method of producing ideas of and in jugaad is one that cannot be separated from the understanding and practice of embodied intuition in different technocultures (Bergson 1988; Galloway and Thacker 2007; Hansen 2004). In such movement-based religious/spiritualist sects such as Buddhism, Sufism, Bhakti Hinduism, Black Methodism, liberation theology, and feminist spiritualism, the practitioners, adherents, converts, and devotees had to work around the norms of the dominant faith and/or political order in order to pursue their own collective emancipation. They borrowed through dominant discourses and institutions, but in ways that refused to take a leap into revolutionary becoming. Or consider emergent aesthetic and ethico-political forms such as ghazal in the nineteenth century, or, in the twentieth, the Hindi film song. In both forms and in the sociocultural life of those forms, the norms of aesthetic creation, value, and expression—the social syntheses of intellectual and manual labor—were challenged through a hybridization of different assemblages of languages, figuration, cultural production, and circulation. These assemblages can be understood as workaround ecologies that literally hack into forms of (religious, aesthetic) power, and they can probably best be situated in the anomalous and unpredictable creation of an ontological and durational transition and intensive flux. Thus, jugaad method follows a minor empiricism:

> Empiricism is often defined as a doctrine according to which the intelligible "comes" from the sensible, everything in the understanding comes from the senses. But that is the standpoint of the history of philosophy: they have the gift of stifling all life in seeking and in positing an abstract first principle. Whenever one believes in a great first principle, one can no longer produce anything but huge sterile dualisms. Philosophers willingly surrender themselves to this and centre their discussions on what should be the first principle (Being, the Ego, the Sensible? . . .). But it is not really worth invoking the concrete richness of the sensible if it is only to make it into an abstract principle. In fact the first principle is always a mask, a simple image. That does not exist, things do not start to move and come alive until the level of the second, third, fourth principle, and these are no longer even principles. Things do not begin to live except in the middle. In this respect what is it that the empiricists found, not in their heads, but in the world, which is like a vital discovery, a certainty of life which, if one really adheres to it, changes one's way of life? It is not the question "Does the intelligible come from the sensible?" but a quite different question, that of relations. *Relations are external to their terms.* "Peter is smaller than Paul," "The glass is on the table": relation is neither internal to one of the terms which would consequently be subject, nor to two together. Moreover, a relation may change without the terms changing. (Deleuze and Parnet 2007, 54–55)

Jugaad is an (un)timely empiricism for a debilitating austerity, its method diagrams the dynamic relations among terms in qualitative and quantitative multiplicities, with gradients of complexity and dimensions of change. Another politics will make that change-phase transition revolutionary.

Jugaad and Rape Cultures in India

At a recent screening and panel discussion of Deepa Mehta's *Anatomy of Violence* (2017), the connection between types of jugaad practice and misogynist rape cultures in India came to the fore. The movie is Mehta's collective and experimental response to the December 16, 2012, gang rape and murder of Jyoti Singh in South Delhi. Singh was a twenty-three-year-old female physiotherapy intern; she was beaten, gang raped, and tortured in a private bus in which she was traveling with her friend, Awindra Pratap Pandey. There were six other men in the bus, including the driver, all of whom raped and assaulted Singh and beat Pandey. Thirteen days after the assault, Singh died from her injuries. All the accused were arrested and charged with sexual assault and murder. One of the accused, Ram Singh, died in police custody from possible suicide. The rest of the accused went on trial in a fast-track court. The juvenile was convicted of rape and murder and given the maximum sentence of three years' imprisonment in a reform facility. On September 10, 2013, the four remaining adult defendants were found guilty of rape and murder and three days later were sentenced to death by hanging. On March 13, 2014, Delhi High Court, in the death reference case and hearing appeals against the conviction by the lower Court, upheld the guilty verdict and the death sentences (see Dutta and Sircar 2013; Lodhia 2015; Roychowdhury 2013).

Mehta's film, according to one film festival that screened it,

> takes a fearless approach to the topic. In collaboration with theatre artist Neelam Mansingh Chowdhry, Mehta worked improvisationally with her actors to envisage possible sociological and psychological backgrounds and pasts for the perpetrators and the victim. The film posits formative events in the men's lives, imagining the origins of their violent, remorseless personalities, while presenting the woman's life in parallel. Unburdened by the weight of a large production, which can sometimes crush an artistic inquiry into the shape of a conventional narrative, Mehta pares things down to the essentials. The film seeks something other than an onscreen trial of these particular individuals. While still holding them accountable, it denounces the patriarchal culture and the cycle of abuse that fed their dark impulses, and the economic system that leaves its disadvantaged classes in desperate straits.[2]

In the course of the panel discussion in London, my colleague Ashwin Devasundaram (see also 2016) noted that "an underclass of homeless people sleep rough in Delhi—some sleep on bridge girders sandwiched between the roadway above and the murky waters below. Economically impoverished men from rural hinterlands in northern Indian states are often forced to migrate to big cities like Delhi. They are then pushed to the margins and often subsist on a bare minimum in slums. Most of the perpetrators had a similar background." He then asked the panelists to consider the push-pull of rural-to-urban migration and the dislocating and alienating effect that it has on these men who often end up as marginalized slum dwellers in Delhi. "Specifically, to what extent does the perpetrators' squalid existence in a Delhi slum act as a canvas to their horrific act?" As one of the panelists, I began by noting that rape culture is trans-class, but has specific class, cultural, and caste articulations, its specific set of ideological and material overdeterminations. The billionaire President Donald Trump is an alleged rapist, for instance. So then if this class cause is removed or radically problematized, how can feminist, queer, postcolonial, and Marxist organizing address the psychobiographies of the working-class perpetrators? In several of the responses to *Anatomy of Violence*, both women and men can be heard to declare (some variant of) "We need to teach men. Men need to feel supported, it needs to be ok, it needs to be a masculine strong thing to be in touch with what you are feeling, with what is going on emotionally and in a man's social relations. Empowered women, but we don't have support structures for men. The lesson here for everyone is NOT to be a bystander, to be a good citizen is to be in active sympathy, in solidarity."[3]

Mehta and Mansingh Chowdhury conducted improvisational workshops with actors, who based their invented stories on the general framework of the perpetrators' data-enriched biographies. Through this method of participatory filmmaking, the film also highlights the relationship between misogynist cultures and subaltern hacking practices. Several of the perpetrators are shown scavenging through the electronic detritus of Delhi's "pirate modernity" (Sundaram 2009), recalling the well-known connection between kabad (recyclable garbage) and jugaad (as refurbishing and repurposing), for example, in the informal cash markets in electronics (e.g., in Nehru Place).[4] The rapists, through assemblages of mobile phone technology and habituation, become a subaltern masculinist "pack" or "group" formed through shared jugaad practices, ranging from substituting SIM cards, to repurposing electronic kabad, or junk, to carjacking. Jugaad in this subaltern ecology becomes not only a way of survival, but also a kind of obsessive orientation toward increasingly individualized gains. What is the connection between jugaad as a method of life hacking and rape cultures in different classes and communities of Indian men? Here the mobile phone becomes literally an instrument of torture. While it would be a gross simplification to say that this correlation with jugaad practice and the decisive aspects of masculinist rape culture is shaped entirely by neoliberal globalization and its attendant social relations, I have rather attempted to diagram the volatile connections that give jugaad practices several very different potential and actual politics at once.

Challenging the panelist for *Anatomy of Violence*, Devasundaram's questions suggested to me that a renewed focus on this question of jugaad media ecologies and rape cultures was necessary. Indeed, the aftermath of Singh's rape and murder emphasized a "clash of two Indias"—one middle-class, aspirational and economically ascendant; the other, feudal, poor, and patriarchal.

This colossal gulf includes not only social, religious, caste-based and economic divisions but also "digital haves and have nots"—can you locate the film's depiction of the perpetrators as being deprived on all these levels—does it gesture to the lack of any form of social security—the absence of a welfare state? What is the relationship between technologies of so-called subaltern development, the post-socialist Indian State, "women's empowerment," and caste and class hierarchies in India? This film, at both the level of creative form and in its strategic montage, sound editing, freeze frames, and general mise en scene of a background of mobile digital image making and technological recycling, poses the question of what it means to narrate, in "relational blocs of duration" (Deleuze 2007, 317–325). There is a history of long-term societal structures complicit with misogynist violence—"the men have their own agency with their own history." How can this history be told today, and how in experimental digital cinema? What are the unspoken complicities performed in the film?

This book has considered gender, sexual, and class and caste struggles for equality and justice in India through different forms of jugaad practice in hacking media ecologies. Throughout, different emancipatory and reactionary articulations of jugaad practice as method, as masculinist and neoliberal forms of capture, have helped to diagram the changing relationship between new technologies of communication and revolutionary becoming in subaltern hacking ecologies. In these newly digitalized ecologies, mobile phone practices blur work and life, intensifying social antagonisms both within and between communities, as well as in the domestic sphere of social reproduction, in which the extraction of value from women's unpaid labor is regularly performed through male violence and by the enforced shared use of the "women's mobile." If there is a philosophy of jugaad, it would be one that attempts a practical experimentation in creatively and collectively reverse-engineering and transvaluating the workaround in the regime of value, sense, and force involved in securing for jugaad a habitual, all too habitual, future.

HOW CAN WE hack the future?

Notes

1. Here the works of Bernard Stiegler, Mark Hansen, Walter Benjamin, Susan Buck-Morss, Martin Heidegger, Jacques Derrida, Gilles Deleuze, Félix Guattari, Gilbert Simondon, Isabelle Stengers, Raniero Panzieri, Manuel DeLanda, Michael Hardt, and Antonio Negri all have important things to add to this argument. Questions of techne and anima, of the archive and the digital, or coevolution of carbon- and silicon-based life—these questions have been layered in the pages that follow, such that their itinerary would form sets of parataxes intercalated between a fetishized status quo of property, security, and territory and a potentially liberatory, ugly, and experimenting undercommons, unbranded and autonomous. This dialectic is not over.

INTRODUCTION

1. See Ajana 2013; Arboleda 2015; Baka 2013; Berardi 2008, 2009b; Bhaskaran 2004; Birkinshaw 2016; Brown 2015; Sen and Dasgupta 2009; Mandarini 2005; Narrain 2008; Streeck 2014; Vanita 2013.

2. Povinelli, drawing on inspiring work by Brian Massumi, pursues the implications of American Pragmatism for the project of decolonizing attention. See her important analysis in *Geontologies* where she discusses the vitalism of the pragmatic frame and its resonances with aboriginal dreaming practices and their associated materialist animism (2016, 30–37, 125–138). Massumi and Povinelli have both informed my own project of decolonizing attention pursued in this study.

3. Contemporary digital marketers are drawing on the work of economist Richard Thaler and Cass Sunstein (1999) to develop ecologies of behavioral change for "better" consumers: "In the wake of economist Richard Thaler's Nobel Prize in Economic Sciences for his research in behavioral economics, staff

writer Hal Conick explores Thaler's revolutionary theories on how to "nudge" people to behave in certain ways and how that theory remains relevant in the digital age. "Perhaps you were nudged by a snack wrapper, imploring you to pick up, unwrap and devour its salty-sweet contents," Conick writes. "Perhaps you were nudged by a mobile notification: respond to a friend request, tip your rideshare driver or—hey, it's raining—order some delivery food." Marketing is, in many ways, a long-game nudge" (Soat 2018, 3). These kinds of nudge campaigns using and abusing jugaad have been fairly consistent in Indian advertising—see for instance the recent makemytrip.com ads featuring Alia Bhatt and Ranveer Singh: https://www.youtube.com/watch?v=n45H7ThwDDs.

4. Robert Hullot-Kentor, Adorno's translator, usefully remarks: "Adorno organized Aesthetic Theory as a paratactical presentation of aesthetic concepts that, by eschewing subordinating structures, breaks them away from their systematic philosophical intention so that the self-relinquishment that is implicit in identity could be critically explicated as what is nonintentional in them: the primacy of the object" (Adorno 2013, xiv).

5. The campaign can be seen at https://www.youtube.com/watch?v=9V4GN _Of6nE.

6. See Udapa 2017; the moralistic anti-jugaad film *Jugaad*; and the criminality attached to jugaadus in *Ishquiyan*. See as well the films *99* and *Black Wednesday*— in which the mobile is linked to various kinds of illegal and antinational activities. All point to a deep popular ambivalence regarding jugaad practice.

7. Snehojit Khan, "Sulekha urges users to go anti-jugaad with its new campaign," afaqs!, January 12, 2016, http://www.afaqs.com/news/story /46816_Sulekha-urges-users-to-go-anti-Jugaad-with-its-new-campaign.

8. Of course, one way to disrupt the drama of habituated discourses is to signify in another: I will often, throughout this study, refer to jugaads as (life) hacking, workarounds, tricks, cons, or reflowing. I have no doubt that I have not touched on the entire range of meanings and senses of this protean term. But I suppose diagrammatic affect plays in the fuzzy set of impossible definitions; jugaad is a term the exact definition of which is less important than the material and psychic relations it mobilizes. This is not, to be clear, a mysticism of the term but an insistence on the slow learning necessary to create common notions. Thus we proceed through hesitations and throat clearings, in short, through paratactical becoming.

9. Jai Arjun Singh, "Cinema and the underdog," *Caravan*, December 1, 2011, http://www.caravanmagazine.in/reviews-and-essays/cinema-and-underdog #sthash.1ryKOlUb.dpuf; "On a documentary titled *Videokaaran*, and Its memorable 'hero,'" *Jabberwock*, September 9, 2011, http://jaiarjun.blogspot.co.uk/2011 /09/on-documentary-titled-videokaaran-and.html).

10. Sagai Raj's casual sexism resonates very differently today after the many gang rapes that have been publicized since 2011. Deepa Mehta's *Anatomy of Vio-*

lence (2016) is a searing commentary on both women's agency in India and the mediatization of rape today.

11. Interview with Sikh taxi driver, Coventry, UK, June 2016.

12. See D-cent, https://dcentproject.eu/; https://www.facebook.com/Why-Loiter-193556873988115/?fref=ts; Phadke et al. 2009; chapter 3 in this book.

13. I will also refer throughout *Jugaad Time* to various texts on Guattari's diagrammatics. Some representative sections follow here. "Guattari locates the emergence of the modern militant aggregation in what he calls the 'Leninist breakthrough' during the 1903 Second Congress of the All-Russian Social Democratic Labour Party, from where—following certain procedural and organisational disputes—emerged a set of affective, linguistic, tactical and organisational traits that constitute a kind of Leninist diagram or abstract machine (Guattari 1984, 184–195). This militant machine, Guattari argues, is characterized by the production of a field of inertia that restricts openness and encourages uncritical acceptance of slogans and doctrine; the hardening of situated statements into universal dogma; the attribution of a messianic vocation to the party; and a domineering and contemptuous attitude—'that hateful "love" of the militant'—to those known as 'the masses' (Guattari 1984, 130). Guattari sees the break of 1903 as the moment that a particular militant diagram was set forth: 'From this fundamental breach, then, the Leninist machine was launched on its career; history was still to give it a face and a substance, but its fundamental encoding, so to say, was already determined' (Guattari 1984, 130). As with any diagram, it draws together its substance in varying ways over time and space, but there is a certain regularity of functions upon which (at least in the 1980s) 'our thinking is still largely dependent today' (Guattari 1984, 190). In discussing the post-'68 French groupuscule milieu Guattari thus contends that the range of groups from anarchist to Maoist may at once be 'radically opposed in their *style:* the definition of the leader, of propaganda, a conception of discipline, loyalty, modesty, and the asceticism of the militant,' but they essentially perform the same militant function of 'stacking,' 'sifting' and 'crushing' desiring energies (Guattari 1995, 59)" (Thoburn 2008, 110). "The ethical wager is to multiply 'existential shifters' to infinity, joining creative mutant Universes. The ontological pragmatic corresponds to this function of existentialization, detecting intensive indices, diagrammatic operators in any point or domain whatever, without any ambition to universalize them, so that what is demanded are not instruments of interpretation but cartographic tools. Even the little 'a' of Lacan, with its admirable deterritorializing character, or the partial objects of Melanie Klein, can be considered as 'crystals of singularization,' 'points of bifurcation outside of dominant coordinates, from which mutant universes of reference might emerge'" (Pelbart 2011, 76). "As a modern philosopher Bergson is novel not because he does not accept the restrictions placed on the philosophy of nature or life by Kant. His originality resides in the manner in which he resists Kant. The con-

ceptions of homogeneous space and time that characterize modern thought are neither properties of things nor essential conditions of our knowledge of them. Rather, they articulate what Bergson calls the 'double work of solidification and division' that we effect on the world—'the moving continuity of the real'—as a means of obtaining a fulcrum for our action: 'They are the diagrammatic design of our eventual action upon matter.' Like Hegel, Bergson makes the charge that Kant's Copernican revolution has the effect of making matter and spirit unknowable. Navigating a way through and beyond the poles of metaphysical dogmatism (whether mechanism or dynamism) and critical philosophy becomes necessary in order to demonstrate that the 'interest' of space and time is not 'speculative' but *vital*. This is why Deleuze insists that it was important to Bergson to demonstrate the entirely empirical character of the *élan vital*, that is, as something that is lived. It will then become possible to gain an insight into the germinal character of life in which the separation between things, objects, and environments is neither absolutely definite nor clear-cut, for 'the close solidarity which binds all the objects of the material universe, the perpetuality of their reciprocal actions and reactions, is sufficient to prove that they have not the precise limits which we attribute to them'" (Ansell-Pearson 2001, 33–34).

FABLES OF THE REINVENTION I

1. All cell phones, as well as smartphones to a greater degree, give off forms of nonionizing electromagnetic radiation called radio frequency (RF) radiation and extremely low frequency (ELF) radiation. RF radiation consists of the cell signal, Bluetooth and Wi-Fi, while ELF radiation is generated by the phone's hardware. This radiation is absorbed into the body, usually through body tissue situated at or near where the cell phone is held. The degree of exposure will depend on several factors, including the type of cell phone being used, how far the user is from the cell phone's antenna, how much time is spent on the cell phone, and how far the user is from cell towers (https://defendershield.com/do -cell-phones-emit-radiation-actually-harmful/).

1. THE AFFECT OF JUGAAD

1. "Acche Din" (good days or times) and "India Shining" are both populist catch phrases of the Bharatiya Janata Party (BJP), India's right-wing party controlled by the Rashtriya Swayamsevak Sangh (RSS; "National Volunteer Organization"), founded in 1925 by Keshav Baliram Hedgewar (1889–1940). The RSS is a proto-fascist and paramilitary organization whose political face is the BJP.

2. According to Wikipedia, jugaad refers to "a creative idea, a quick, alternate way of solving or fixing problems"; colloquially it means a quick workaround that overcomes commercial, logistical, or legal obstacles. Derived from the Pun-

jabi word for a makeshift jalopy or "clunker" that will just barely get from point a to b, jugaad has become a kind of social enterprise movement gathering together "a community of enthusiasts, believing it to be the proof of Indian bubbling creativity, or a cost-effective way to solve the issues of everyday life" (*Wikipedia* 2018); see also *Oxford Living Dictionary* 2018: "A flexible approach to problem-solving that uses limited resources in an innovative way. 'Countries around the world are beginning to adopt jugaad in order to maximize resources' as modifier 'jugaad entrepreneurs.'" Extending these definitions through PDMA discourse, jugaad innovators can be defined as "lead users": "Users for whom finding a solution to one of their consumer needs is so important that they have modified a current product or invented a new product to solve the need themselves because they have not found a supplier who can solve it for them. When these consumers' needs are portents of needs that the center of the market will have in the future, their solutions are new product opportunities" (Kahn 2005, 593; see also http://ennovient.com/blogs/).

3. "History of Indian Telecommunications," telecomtalk.info, accessed January 2, 2014.

4. Airtel-related information is from http://www.airtel.in, accessed September 12, 2013.

5. For examples, see *Three Idiots* (2009; dir. R. Hirani), *Style* (2001; dir. N. Chandra), *College Girl* (1960; dir. T. P. Rao), and *Main Hoon Na* (2004; dir. F. Khan).

6. I owe this phrasing regarding resistance, and the immeasurable benefit of a sustained and convivial exploration of its historiography, to Stefano Harney.

2. NEOLIBERAL ASSEMBLAGES OF PERCEPTION

1. In A. N. Whitehead's theory (1979) of perception of actual entities, ingression is the objectification of one actual entity through the prehensions of another actual entity.

2. See Dipesh Chakrabarty's (2014) analysis of the Anthropocene; Jason Moore's critique of the web of life under capital; Jodi Dean's (2012), Michael Hardt's (1999), and David Harvey's (2012) recent works on the radical critique of environmentalism are also instructive in moving beyond a human-centric approach to ecologies of sensation.

3. Interview conducted by Rachna Kumar, January 2017. Names have been anonymized. In the interview, Bombay is used for Mumbai.

4. CIDCO is the City and Industrial Development Corporation of Maharashtra Ltd., and it played a role in developing the social and physical infrastructure of Navi Mumbai in 1971. CIDCO was set up by the Government of Maharashtra (GoM) as a public limited company under the Indian Companies Act and it is wholly owned by the GoM. In March 1971, CIDCO was designated as the New

Town Development Authority for Navi Mumbai. The seed capital given to CIDCO [City and Industrial Development Corporation] was Rs. 3.95 cr. and it was expected that it would use land as a resource to finance the project (https://cidco.maharashtra.gov.in/#).

5. Hardt goes on to note that the term "service" covers a large range of activities from health care, education, and finance, to transportation, entertainment, and advertising. The jobs, for the most part, are highly mobile and involve flexible skills. More important, they are characterized in general by the central role played by knowledge, information, communication, and affect. In this sense, we can call the neoliberal economy an informational economy (Hardt 1999, 91).

6. According to NASSCOM (National Association of Software and Services Companies), in fiscal year 2014, India's information technology and business process management industry would add $12–15 billion incremental revenue to existing industry revenues of $118 billion. During fiscal year 2014, industry's exports are estimated to grow 13 percent to $86 billion, with domestic revenues up 9.7 percent at Rs. 1.910 billion. NASSCOM reports also stated that the industry added 160,000 employees in 2013 and had provided direct employment to 3.1 million people and indirect employment to 10 million people. Exports by India's IT outsourcing sector are expected to rise 13–15 percent in the fiscal year starting April 2014, as an improving global economy encourages banks and companies to boost spending on technology. NASSCOM has forecasted IT services exports in 2014–2015 to rise to as much $99 billion. The increase in growth rate compares with an estimated 13 percent rise in fiscal year 2016. By 2017–2018, yearly growth for BSO was over 20 percent a year (see https://www.business-standard.com/article/news-cm/india-s-services-exports-rises-20-3-in-february-2018-118041400355_1.html, accessed July 21, 2018). It also states that the Indian IT and ITeS (information technology enabled services) industry is likely to grow to about $300 billion by 2020, focusing on areas like e-commerce, software products, and the IT market (http://info.shine.com/industry/it-ites-bpo/11.html).

7. See Whitson Gordon, "Everything you need to know about rooting your Android phone," September 4, 2013, http://lifehacker.com/5789397/the-always-up-to-date-guide-to-rooting-any-android-phone.

3. JUGAAD ECOLOGIES OF SOCIAL REPRODUCTION

1. This analysis is informed by the Italian autonomist tradition, especially those associated with the compositionist tendency—such as Hardt and Negri, Paolo Virno, Maurizio Lazzarato, and Franco Berardi—which sees biopolitical resistance and subjectivity as an autonomous power capable of creating alternative forms of life and new patterns of relations between human and nonhuman worlds (Arboleda 2015, 39–40; Berardi 2006; Hardt and Negri 2001; Hardt and Negri 2009, 211; Lazzarato 2006; Virno 2003).

2. A term originally coined by Marshal Sahlins, the domestic mode of production, in the words of Paltasingh and Lingam, includes "that system of household labour in which the household members produce use values for direct consumption or accumulation within the household. The forms of exploitation of the domestic mode of production is based in the labour of female 'dependents' within the household and because the male head(s) of household is expropriating surplus labour when he consumes the use values produced by his dependents. He benefits from this relation of exploitation, both in the use values he appropriates and on the leisure time resulting from the necessary labour time he relinquishes" (Paltasingh and Lingam 2014, 46; Safri and Graham 2010; Sahlins 1972).

3. Interview conducted in Hindi by Anisha Saigal, June 4, 2015, in South Delhi. Name anonymized.

4. This case is corroborated by my own fieldwork in Delhi, Bangalore, Bhopal, and Mumbai from 2009 to 2014 in mobile phone culture and kinship relations: again and again, men expressed anxiety that their control over women in their family was compromised by the "openness" of the mobile phone.

5. The changing struggle in and through this assemblage is affected by another important parameter: digital memory. The amount of memory affects directly the qualitative experience of media consumption through the device. Rekha senses she has a lot of media on the phone (her shared phone has a 2 GB memory card). "There must be thirty to thirty-five videos we recorded and approximately the same number of audio tracks. Besides this, we have photographs. The memory card is not full." The amount of available memory, for instance, affects the lived and virtual experience of social events or community "functions, when we use the mobile camera to click pictures, make videos."

6. Interview conducted in Hindi by Anisha Saigal, July 10, 2015. Name anonymized.

4. DIAGRAMMING AFFECT

1. "Mobile Vas in India: 2010," June 2010, Internet and Mobile Association of India, http://www.iamai.in/Upload/Research/Report_on_MVAS_(2010)_submittal_42.pdf, p. 31.

2. See "Mobile Vas in India: 2010," June 2010, Internet and Mobile Association of India (IMAI 2010), https://business.mapsofindia.com/communications-industry/internet/association.html.

3. See, for instance, http://jscljabalpur.org/.

4. Interview conducted by Anisha Saigal, March 15, 2015.

5. See, for instance, in Mumbai, the work of Akshara (http://www.akshara centre.org/reclaimingcity/), Majlis (http://www.majlislaw.com/), and Point of View (http://pointofview.org/).

6. See Neha Singh, "Let's start a revolution . . . but how?," *Why Loiter?*, http://whyloiter.blogspot.co.uk/.

CONCLUSION

1. See Ahmed 2007; Berardi 2008; Berlant 2016; Butler 2006; Clough 2010; De Angelis 2007, 2017; Hardt and Negri 1994, 2001, 2009; Mazzadra 2011; Precarious Workers Brigade 2015; Puar 2017; Roy 2009; Spivak 1999.

2. Toronto International Film Festival, http://www.tiff.net/films/anatomy-of-violence/.

3. See the clip of the Deepa Mehta panel, https://www.youtube.com/watch?v=8qV4karzV8I.

4. See, for example, the "Kabad to Jagaad" project, https://www.youtube.com/watch?v=HKjumXXFJ9g; and the "jugaad baaz" characters in Bollywood B films, such as *99* (2009, dir. Krishna D.K. and Raj Nidimoru) and *Fukrey* (2013, dir. Lamba).

References

Adams, R. E. 2014. "Natura urbans, natura urbanata: Ecological urbanism, circulation, and the immunization of nature." *Environment and Planning D: Society and Space* 32(1): 12–29.

Adams, V., M. Murphy, and A. E. Clarke. 2009. "Anticipation: Technoscience, life, affect, temporality." *Subjectivity* 28(1): 246–265.

Adorno, T. W. 2013. *Aesthetic theory*. London: Continuum Books.

Agamben, G. 2005. *State of exception*. Translated by K. Attell. Chicago: University of Chicago Press, 2005.

Aggarwal, M., and V. Gupta. 2009. "Comparative study of telecom service providers in India." Proceedings of the 2009 IEEE Systems and Information Engineering Design Symposium, University of Virginia, Charlottesville, April 24, 2009, 107–112.

Ahmed, S. 2007. *The cultural politics of emotion*. London: Routledge.

Ajana, B. 2013. *Governing through biometrics: The biopolitics of identity*. London: Macmillan Palgrave.

Amin, A. 2009. "Locating the social economy." In *The social economy, international perspectives on economic solidarity*, edited by A. Amin, 3–21. New York: Zed Books.

Amin, A., and N. Thrift. 2002. *Cities: Reimagining the urban*. New York: Polity.

Amoore, L. 2013. *The politics of possibility: Risk and security beyond probability*. Durham, NC: Duke University Press.

Anand, D. 2011. *Hindu nationalism in India and the politics of fear*. New York: Palgrave Macmillan.

Anantharaja, A. 2009. "Causes of attrition in BPO companies: Study of a mid-size organization in India." *IUP Journal of Management Research* 8(11): 13–27.

Anatomy of Violence. 2016. Directed by D. Mehta. Hamilton Mehta Productions. 93 min. Canada.

Anderson, B. 2004. "Time-stilled space-slowed: How boredom matters." *Geoforum* 35: 739–754.

Anderson, B. 2005. "Practices of judgment and domestic geographies of affect." *Social and Cultural Geography* 6: 645–660.

Anderson, B. 2007. "Hope for nanotechnology: Anticipatory knowledge and governance of affect." *Area* 19: 156–165.

Anderson, B., and J. Wylie. 2009. "On geography and materiality." *Environment and Planning A* 41: 318–335.

Andres, L., and C. Chapain. 2013. "The integration of cultural and creative industries into local and regional development strategies in Birmingham and Marseille: Towards an inclusive and collaborative governance?" *Regional Studies* 47(2): 161–182.

Aneesh, A. 2001. "Virtual migrations: Indian programmers in the US-based information industry." PhD diss. New Brunswick, NJ: Rutgers University Press.

Aneesh, A. 2006. *Virtual migration: The programming of globalization.* Durham, NC: Duke University Press.

Anon. 2016. "Ogilvy & Mather India urges users to 'Go #AntiJugaad' with Sulekha local services app." CampaignBriefAsia.com. Accessed May 1, 2018. http://www.campaignbriefasia.com/2016/01/ogilvy-mather-india-urges-user.html.

Ansell-Pearson, K. 2001. *Philosophy and the adventure of the virtual: Bergson and the time of life.* London: Routledge.

Ansems de Vries, L., and D. Rosenow. 2015. "Opposing the opposition? Binarity and complexity in political resistance." *Environment and Planning D: Society and Space* 33(6): 1118–1134.

Arboleda, M. 2015. "The biopolitical production of the city: Urban political ecology in the age of immaterial labour." *Environment and Planning D: Society and Space* 33: 35–51.

Arora, A., and S. Athreye. 2002. 'The software industry and India's economic development.' *Information Economics and Policy* 14(2): 253–273.

Arora, P., and N. Rangaswamy. 2013. "Digital leisure for development: Reframing new media practice in the Global South." *Media, Culture and Society* 35(7): 898–905.

Asad, T., ed. 1975. *Anthropology and the colonial encounter.* New York: Prometheus Books.

Ash, J. 2010. "Architectures of affect: Anticipating and manipulating the event in processes of videogame design and testing." *Environment and Planning D* 28: 653–671.

Ash, J. 2012. "Technology, technicity, and emerging practices of temporal sensitivity in videogames." *Environment and Planning A* 44(1): 187–203.

Ash, J. 2013. "Rethinking affective atmospheres: Technology, perturbation and space times of the non-human." *Geoforum* 49: 20–28.

Ash, J. 2014. "Technology and affect: Towards a theory of inorganically organised objects." *Emotion, Space and Society.* Accessed July 22, 2015. http://dx.doi.org/10.1016/j.emospa.2013.12.017.

Ash, J. 2015a. "Sensation, networks, and the GIF: Toward an allotropic account of affect." In *Networked affect*, edited by Ken Hillis, Susanna Paasonen, and Michael Petit. 119–113. Cambridge, MA: MIT Press.

Ash, J. 2015b. "Technology and affect: Towards a theory of inorganically organized objects." *Emotion, Space and Society* 14: 84–90.

Ashcraft, K. L. 2017. "Submission to the rule of excellence: Ordinary affect and precarious resistance in the labor of organization and management studies." *Organization* 24(1): 36–58.

Athique, A., V. Parthasarathi, and S. Srinivas, eds. 2018. *The Indian media economy*, vol. 1: *Industrial dynamics and cultural adaptation*. Delhi: Oxford University Press.

Baines, P., C. Fill, and K. Page. 2011. *Marketing*. Oxford: Oxford University Press.

Bains, P. 2006. *The primacy of semiosis: An ontology of relations*. Toronto: University of Toronto Press.

Baka, J. 2013. "The political construction of wasteland: Governmentality, land acquisition and social inequality in South India." *Development and Change* 44: 409–428.

Balakrishnan, P. 2006. "Benign neglect or strategic intent? Contested lineage of Indian software industry." *Economic and Political Weekly* 41: 3865–3872.

Balsamo, A. M. 1996. *Technologies of the gendered body: Reading cyborg women*. Durham, NC: Duke University Press.

Bannerji, H. 2000. "Projects of hegemony: Towards a critique of subaltern studies' 'resolution of the women's question.'" *Economic and Political Weekly* 35(11): 902–920.

Barad, K. 2007. *Meeting the universe halfway: Quantum physics and the entanglement of matter and meaning*. Durham, NC: Duke University Press.

Barbagallo, C. 2015. "Leaving home: Slavery and the politics of reproduction." *Viewpoint Magazine*, 5. Accessed November 4, 2015. viewpointmag.com.

Barbagallo, C., and S. Federici. 2012. "Introduction to 'care work' and the commons." *The Commoner* 15: 1–21.

Bassnett, S., and H. Trivedi, eds. 1999. *Post-colonial translation: Theory and practice*. London: Routledge.

Bateson, G. 1972. *Steps to an ecology of mind*. Chicago: University of Chicago Press.

Batty, M. 1995. "The computable city." *International Planning Studies* 2: 155–173.

Beer, D. 2013. *Popular culture and new media: The politics of circulation*. Basingstoke, UK: Palgrave Macmillan.

Beer, D. 2015. "Productive measures: Culture and measurement in the context of everyday neoliberalism." *Big Data and Society* 2: 1–12.

Beneria, L. 1979. "Reproduction, production and the sexual division of labour." *Cambridge Journal of Economics* 3(3): 203–225.

Beneria, L. 1982. "Accounting for women's work." In *Women and development:*

The sexual division of labour in developing societies, edited by L. Beneria, 119–147. New York: Praeger.

Beneria, L., and G. Sen. 1981. "Accumulation, reproduction and women's role in economic development: Boserup revisited." *Signs* 7(2): 279–298.

Benjamin, W. 1999. *The arcades project*. Translated by R. Tiedemann. Cambridge, MA: Belknap Press.

Benjamin, W. 2011. *Illuminations*. New York: Vintage.

Bennett, J. 2010. *Vibrant matter: A political ecology of things*. Durham, NC: Duke University Press.

Berardi, F. 2008. *Félix Guattari: Thought, friendship and visionary cartography*. Translated by G. Mecchia and C. J. Stivale. New York: Palgrave Macmillan.

Berardi, F. 2009a. *Precarious rhapsody: Semiocapitalism and the pathologies of the post-alpha generation*. Translated by A. Bove, E. Empson, M. Goddard, G. Mecchia, A. Schintu, and S. Wright. London: Minor Compositions.

Berardi, F. 2009b. *The Soul at work: From alienation to autonomy*. Los Angeles: Semiotext(e).

Bergson, H. 1988. *Matter and memory*. New York: Zone Books.

Bergson, H. 2012. *Creative evolution*. New York: Dover Publications.

Berlant, L. 2016. "The commons: Infrastructure for troubling times." *Environment and Planning D: Society and Space* 34(3): 393–419.

Berndt, C. and M. Boeckler. 2011. "Performative regional (dis) integration: transnational markets, mobile commodities, and bordered North-South differences." *Environment and Planning A* 43(5): 1057–1078.

Besserud, K., M. Sarkisian, P. Enquist, and C. Hartman. 2013. "Scales of metabolic flows: Regional, urban and building systems design at SOM." *Architectural Design* 83(4): 86–93.

Bhabha, H. 1984. "Signs taken for wonders." In *Europe and its others: Proceedings of the Essex Conference on the Sociology of Literature*, edited by F. Barker, vol. 1, 89–105. Colchester, UK: University of Essex.

Bhabha, H. 1994. *The location of culture*. London: Routledge.

Bhaskaran, S. 2004. *Made in India: Decolonizations, queer sexualities, trans/national projects*. New York: SpringerPublications.

Bhavnani, A., R. Won-Wai Chiu, S. Janakiram, and P. Silarszky. 2008. *The role of mobile phones in sustainable rural poverty reduction: World Bank report—ICT policy division*. Accessed October 13, 2017. http://siteresources.worldbank.org/EXTINFORMATIONANDCOMMUNICATIONANDTECHNOLOGIES/Resources/The_Role_of_Mobile_Phones_in_Sustainable_Rural_Poverty_Reduction_June_2008.pdf.

Birkenholtz, T. 2010. "'Full-cost recovery': Producing differentiated water collection practices and responses to centralized water networks in Jaipur, India." *Environment and Planning A* 42(9): 2238–2253.

Birkinshaw, M. 2016. "Politics, information technology and informal infrastructures in urban governance." *Economic and Political Weekly* 51(5): 57–63.

Birtchnell, T. 2011. "Jugaad as systemic risk and disruptive innovation in India." *Contemporary South Asia* 19(4): 357–372.

Bissell, D. 2008. "Comfortable bodies: Sedentary affects." *Environment and Planning A* 40: 1697–1712.

Bissell, D. 2009. "Obdurate pains, transient intensities: Affect and the chronically pained body." *Environment and Planning A* 41: 911–928.

Bittman, M., J. Brown, and J. Wajcman. 2009. "The mobile phone, perpetual contact and time pressure." *Work, Employment and Society* 23(4): 673–691.

Blackburn, S. 2014. "The politics of scale and disaster risk governance: Barriers to decentralisation in Portland, Jamaica." *Geoforum* 52: 101–112.

Blackman, L. 2008. "Affect, relationality and the 'problem of personality.'" *Theory, Culture and Society* 25(1): 23–47.

Bloomfield, B., and K. Dale. 2015. "Fit for work? Redefining 'normal' and 'extreme' through human enhancement technologies." *Organization* 22(4): 552–569.

Boddy, C., D. Miles, C. Sanyal, and M. Hartog. 2015. "Extreme managers, extreme workplaces: Capitalism, organizations and corporate psychopaths." *Organization* 22(4): 530–551.

Bogost, I. 2006. *Unit operations: An approach to videogame criticism.* Cambridge, MA: MIT Press.

Bohm, D. 1980. *Wholeness and the implicate order.* London: Routledge.

Boltanski, L., and E. Chiapello. 2007. *The new spirit of capitalism.* London: Verso.

Borsch, C., K. B. Hansen, and A. Lange, A. 2015. "Markets, bodies, and rhythms: A rhythmanalysis of financial markets from open-outcry trading to high-frequency trading." *Environment and Planning D: Society and Space* 33(6): 1080–1097.

Bourdieu, P. 1976. *Marriage strategies as strategies of social reproduction.* Baltimore: Johns Hopkins University Press.

Bourdieu, P. 1984. *Distinction: A social critique of the judgement of taste.* Cambridge, MA: Harvard University Press.

Bourdieu, P. 1996. *The rules of art: Genesis and structure of the literary field.* Stanford, CA: Stanford University Press.

Braun, B. 2014. "A new urban dispositif? Governing life in an age of climate change." *Environment and Planning D: Society and Space* 32(1): 49–64.

Braun, B., and S. Wakefield. 2014. Guest editorial. *Environment and Planning D: Society and Space* 32(1): 4–11.

Brosius, C., and M. Butcher. 1999. *Image journeys: Audio-visual media and cultural change in India.* New Delhi: Sage.

Brown, W. 2015. *Undoing the demos: Neoliberalism's stealth revolution*. New York: Zone Books.

Buck-Morss, S. 1991. *The dialectics of seeing: Walter Benjamin and the Arcades Project*. Cambridge, MA: MIT Press.

Burton, A. 1997. "House/daughter/nation: Interiority, architecture, and historical imagination in Janaki Majumdar's 'Family History.'" *Journal of Asian Studies* 56(4): 921–946.

Butler, J. 2006. *Precarious life: The powers of mourning and violence*. London: Verso.

Caffentzis, G. 2005. "Immeasurable value? An essay on Marx's legacy." *The Commoner* 10: 87–114.

Cameron, J., and K. Gibson. 2005. "Alternative pathways to community and economic development: The Latrobe Valley Community Partnering Project." *Geographical Research* 43(3): 274–285.

Camfield, D. 2007. "The multitude and the kangaroo: A critique of Hardt and Negri's theory of immaterial labour." *Historical Materialism* 15(2): 21–52.

Caragliu, A., C. D. Bo, and P. Nijkamp. 2011. "Smart cities in Europe." *Journal of Urban Technology* 18: 65–82.

Carswell, G., and G. De Neve. 2014. "T-shirts and tumblers: Caste, dependency and work under neoliberalisation in South India." *Contributions to Indian Sociology* 48: 103–131.

Castells, M. 2006. *Mobile communication and society: A global perspective*. Cambridge, MA: MIT Press.

Castells, M. 2015. *Networks of outrage and hope: Social movements in the internet age*. Cambridge, UK: Polity.

Chakrabarty, D. 2009. *Provincializing Europe: Postcolonial thought and historical difference*. Princeton, NJ: Princeton University Press.

Chakrabarty, D. 2014. "Climate and capital: On conjoined histories." *Critical Inquiry* 41(1): 1–23.

Chandra, N. K. 2009. "China and India: Convergence in economic growth and social tensions?" *Economic and Political Weekly*: 41–53.

Changeux, J. 2004. *The physiology of truth: Neuroscience and human knowledge*. Cambridge, MA: Belknap Press.

Chatterjee, P. 1989. "Colonialism, nationalism, and colonialized women: The contest in India." *American Ethnologist* 16(4): 622–633.

Chatterjee, P. 1990. "The nationalist resolution of the women's question." In *Recasting women: Essays in Indian colonial history*, edited by K. Sangari and S. Vaid, 233–253. New Brunswick, NJ: Rutgers University Press.

Chatterjee, P. 1995. *The nation and its fragments: Colonial and postcolonial histories*. New Delhi: Oxford University Press.

Chen, J. E., T. H. Ouyang, and S. L. Pan. 2013. "The role of feedback in changing organizational routine: A case study of Haier, China." *International Journal of Information Management* 33(6): 971–974.

Cheyfitz, E. 1991. *The poetics of imperialism: Translation and Colonization.* Philadelphia: University of Pennsylvania Press.

Clark, A. 2003. *Natural-born cyborgs: Why minds and technologies are made to merge.* New York: Oxford University Press.

Clark, N. 2017. "Politics of strata." *Theory, Culture and Society* 34(2–3): 211–231.

Clough, P. T. 2010. "Afterword: The future of affect studies." *Body and Society* 16: 222–230.

Clough, P. T. 2018. *The user unconscious: On affect, media, and measure.* Minneapolis: University of Minnesota Press.

Clough, P. T., and J. Halley, eds. 2007. *The affective turn.* Durham, NC: Duke University Press.

Cohn, B. S. 1987. *An anthropologist among the historians and other essays.* New Delhi: Oxford University Press.

Coleman, E. G., and A. Golub. 2008. "Hacker practice: Moral genres and the cultural articulation of liberalism." *Anthropological Theory* 8(3): 255–277.

Combes, M. 2013. *Gilbert Simondon and the philosophy of the transindividual.* Cambridge, MA: MIT Press.

Connolly, W. E. 2002. *Neuropolitics: Thinking, culture, speed.* Minneapolis: University of Minnesota Press.

Cowen, D., and A. Siciliano. 2011. "Surplus masculinities and security." *Antipode* 43(5): 1516–1541.

Culp, A. 2016. *Dark Deleuze.* Minnesota: University of Minneapolis Press.

Dale, K., and Y. Latham. 2015. "Ethics and entangled embodiment: Bodies—materialities—organization." *Organization* 22(2): 166–182.

Dalla Costa, M. 2015. "Introduction to the archive of the feminist struggle for wages for housework." *Viewpoint Magazine,* 5. Accessed November 4, 2015. viewpointmag.com.

Dasgupta, R. K., and D. Dasgupta. 2018. *Queering digital India: Activisms, identities, subjectivities.* Edinburgh: Edinburgh University Press.

Dastur, F. 2000. "Phenomenology of the event: Waiting and surprise." *Hypatia* 15(4): 178–189.

Dean, J. 2009. *Democracy and other neoliberal fantasies: Communicative capitalism and left politics.* Durham, NC: Duke University Press.

Dean, J. 2012. *The communist horizon.* London: Verso.

De Angelis, M. 2003. "Reflections on alternatives, commons and communities or building a new world from the bottom up." *The Commoner* 6 (Winter): 1–14. Accessed November 4, 2015. http://www.thecommoner.org.

De Angelis, M. 2007. *The beginning of history: Value struggles and global capital.* London: Pluto Press.

De Angelis, M. 2010. "The production of commons and the 'explosion' of the middle class." *Antipode* 42(4): 954–977.

De Angelis, M. 2017. *Omnia sunt communia on the commons and the transformation to postcapitalism*. London: Zed Books.

De Certeau, M. 1984. *The practice of everyday life*. Berkeley: University of California Press, 1988.

DeLanda, M. 1997. *A thousand years of nonlinear history*. New York: Zone Books.

DeLanda, M. 2002. *Intensive science and virtual philosophy*. New York: Continuum.

DeLanda, M. 2010. *Philosophy and simulation: The emergence of synthetic reason*. London: Continuum.

Deleuze, G. 1986. *Cinema 1: The movement-image*. Translated by Hugh Tomlinson and Barbara Habberjam. Minneapolis: University of Minnesota Press.

Deleuze, G. 1988a. *Bergsonism*. Translated by Hugh Tomlinson and Barbara Habberjam. London: MIT Press.

Deleuze, G. 1988b. *Spinoza: Practical philosophy*. San Francisco: City Lights.

Deleuze, G. 1991. *Empiricism and subjectivity: An essay on Hume's theory of human nature*. New York: Columbia University Press.

Deleuze, G. 1992a. *Expressionism in philosophy: Spinoza*. Translated by M. Joughin. New York: Zone Books.

Deleuze, G. 1992b. "Postscript on the societies of control." *October* 59: 3–7.

Deleuze, G. 1994. *Difference and repetition*. New York: Columbia University Press.

Deleuze, G. 1998. *Essays critical and clinical*. New York: Verso.

Deleuze, G. 2007. *Two regimes of madness: Texts and interviews 1975–1995*. Translated by Ames Hodges and Mike Taormina. New York: Semiotext(e).

Deleuze, G., and F. Guattari. 1983. *Anti-Oedipus: Capitalism and schizophrenia*. Minneapolis: University of Minnesota Press.

Deleuze, G., and F. Guattari. 1987. *A thousand plateaus*. Translated by B. Massumi. New York: Continuum.

Deleuze, G., and F. Guattari. 1994. *What is philosophy?* Translated by H. Tomlinson and G. Burchell. New York: Columbia University Press.

Deleuze, G., and C. Parnet. 2007. *Dialogues II*. New York: Columbia University Press.

Derrida, J. 1998. *Archive fever: A Freudian impression*. Chicago: University of Chicago Press.

Derrida, J. 2016. *Of grammatology*. Baltimore: Johns Hopkins University Press.

Derrida, J., and B. Stiegler. 2002. *Echographies of television*. Cambridge, UK: Polity.

de Souza e Silva, A. 2006. "From cyber to hybrid: Mobile technologies as interfaces of hybrid spaces." *Space and Culture* 9(3): 261–278. doi: 10.1177/1206331206289022.

Devasundaram, A. I. 2016. *India's new independent cinema: Rise of the hybrid*. New York: Routledge.

Dewsbury, J.-D. 2000. "Performativity and the event: Enacting a philosophy of difference." *Environment and Planning D: Society and Space* 18: 473–496.

Dewsbury, J.-D. 2003. "Witnessing space: 'Knowledge without contemplation.'" *Environment and Planning A* 35(11): 1907–1932.

Dewsbury, J.-D. 2012. "Affective habit ecologies: Material dispositions and immanent inhabitations." *Performance Research—a Journal of the Performing Arts* 17: 74–82.

Dey, I., and G. Grappi. 2015. "Beyond zoning: India's corridors of 'development' and new frontiers of capital." *South Atlantic Quarterly* 114: 153–170.

Dowling, E. 2007. "Producing the dining experience: Measure, subjectivity and the affective worker." *Ephemera* 7: 117–132.

Dowling, E., R. Nunes, and B. Trott. 2007. "Immaterial and affective labour explored." *Ephemera* 7: 1–7.

Downey, L. 2013. "The interruption: Investigating subjectivity and affect." *Environment and Planning D: Society and Space* 31: 628–644.

Droege, P., ed. 1997. *Intelligent environments: Spatial aspects of the information revolution*. Amsterdam: Elsevier.

Duberley, J., M. Carrigan, J. Ferreira, and C. Bosangit. 2017. "Diamonds are a girl's best friend . . . ? Examining gender and careers in the jewellery industry." *Organization* 24(3): 355–376.

Duff, C. 2010. "On the role of affect and practice in the production of place." *Environment and Planning D: Society and Space* 28: 881–895.

Dutta, D., and O. Sircar. 2013. "India's winter of discontent: Some feminist dilemmas in the wake of a rape." *Feminist Studies* 39(1): 293–306.

Easterling, K. 2014. *Extrastatecraft: The power of infrastructure space*. New York: Verso Books.

Edensor, T. 2010. "Walking in rhythms: place, regulation, style and the flow of experience." *Visual Studies* 25(1): 69–79.

Edkins, T. 2011. *Wofe*. Performance. Department of Theatre, Queen Mary, University of London.

Ekers, M., and A. Loftus. 2012. "Revitalizing the production of nature thesis: A Gramscian turn?" *Progress in Human Geography* 37(2): 1–19.

Engels, F. 1940. *The origins of the family, private property and the state*. London: Lawrence and Wishart.

Ettlinger, N. 2016. "The governance of crowdsourcing: Rationalities of the new exploitation." *Environment and Planning A* 48: 2162–2180.

Federici, S. 2003. *Caliban and the witch: Women, the body and primitive accumulation*. New York: Autonomedia.

Federici, S., C. G. Caffentzis, and O. Alidou. 2000. *A thousand flowers: Social struggles against structural adjustment in African universities*. Trenton, NJ: Africa World Press.

Fleming, D., and D. Sturm. 2011. *Media, masculinities and the machine:*

F1, transformers and fantasizing technology at its limits. New York: Continuum.

Fleming, P., and A. Sturdy. 2009. "'Just be yourself!': Towards neo-normative control in organisations?" *Employee Relations* 31(6): 569–583.

Florida, R. 2012. *The rise of the creative class*. New York: Basic Books.

Florida, R. 2017. *The new urban crisis: How our cities are increasing inequality, deepening segregation, and failing the middle class and what we can do about it*. New York: Basic Books.

Fotaki, M., K. Kenny, and S. J. Vachhani. 2017. "Thinking critically about affect in organization studies: Why it matters." *Organization* 24(1): 3–17.

Foucault, M. 1995. *Discipline and punish: The birth of the prison*. New York: Vintage Books.

Foucault, M. 2003. *Society must be defended: Lectures at the Collège de France, 1975–1976*. Translated by D. Macey. London: Picador.

Foucault, M. 2007. *Security, territory, population: Lectures at the Collège de France 1977–1978*. Translated by G. Burchell. Edited by M. Senellart. New York: Palgrave.

Foucault, M. 2008. *The birth of biopolitics: Lectures at the Collège de France, 1978–79*. Translated by M. Senellart. New York: Palgrave Macmillan.

Fraser, B. 2008. "Toward a philosophy of the urban: Henri Lefebvre's uncomfortable application of Bergsonism." *Environment and Planning D: Society and Space* 26(2): 338–358.

Fuller, M. 2004. *Media ecologies*. Cambridge, MA: MIT Press.

Gabrys, J. 2014. "Programming environments: Environmentality and citizen sensing in the smart city." *Environment and Planning D: Society and Space* 32: 30–48.

Galloway, A. R. 2004. *Protocol: How control exists after decentralization*. Cambridge, MA: MIT Press.

Galloway, A. R., and E. Thacker. 2007. *The exploit: A theory of networks*. Minneapolis: University of Minnesota Press.

Gandy, M. 2006. "Zones of indistinction: Bio-political contestations in the urban arena." *Cultural Geographies* 13(4): 497–516.

Gane, N., and D. Beer. 2008. *New media*. New York: Berg.

Ganti, T. 2012. *Producing Bollywood: Inside the contemporary Hindi film industry*. Durham, NC: Duke University Press.

Gascoigne, C., E. Parry, and D. Buchanan. 2015. "Extreme work, gendered work? How extreme jobs and the discourse of 'personal choice' perpetuate gender inequality." *Organization* 22(4): 457–475.

Gates, A. 2006. "Helping U.S. companies export white-collar jobs." *New York Times*, December 26.

Ghertner, D. A. 2014. "India's urban revolution: Geographies of displacement beyond gentrification." *Environment and Planning A* 46: 1554–1571.

Ghosal, S. G. 2005. "Major trends of feminism in India." *Indian Journal of Political Science* 66(4): 793–812.

Gibson-Graham, J. K. 1996. *The end of capitalism (as we knew it): A feminist critique of political economy*. London: Wiley-Blackwell.

Gibson-Graham, J. K. 2006. *A postcapitalist politics*. Minneapolis: University of Minnesota Press.

Gibson-Graham, J. K. 2008. "Diverse economies: Performative practices for 'other worlds.'" *Progress in Human Geography* 32(5): 613–632.

Gill, R., and A. Pratt. 2008. "In the social factory? Immaterial labour, precariousness and cultural work." *Theory, Culture and Society* 25(7–8): 1–30.

Glasmeier, A., and S. Christopherson. 2015. "Thinking about smart cities." *Cambridge Journal of Regions, Economy and Society* 8: 3–12.

Glück, Z. 2015. "Piracy and the production of security space." *Environment and Planning D: Society and Space* 33(4): 642 –659.

GoI (Government of India). 2013. *Right to fair compensation and transparency in land acquisition, rehabilitation and resettlement Act 2013*. New Delhi: Ministry of Law and Justice.

GoI (Government of India). 2014a. *Draft concept note on smart city scheme*. New Delhi: Ministry of Law and Justice.

GoI (Government of India). 2014b. *Draft concept note on smart city scheme* (December). New Delhi: Ministry of Urban Development.

GoI (Government of India). 2015. *Smart cities—mission statement and guidelines*. New Delhi: Ministry of Urban Development.

Gopal, M. 2013. "Sexuality and social reproduction: Reflections from an Indian feminist." *Indian Journal of Gender Studies* 20(2): 235–251.

GoUK. 2016. *India's smart cities programme: The UK offer to build together*. London: UK Trade and Investment.

Gradin, S. 2015. "Radical routes and alternative avenues: How cooperatives can be non-capitalist." *Review of Radical Political Economics* 47(2): 141–158.

Graham, S., and S. Marvin. 2001. *Splintering urbanism: Networked infrastructures, technological mobilities and the urban condition*. London: Routledge.

Granter, E., L. McCann, and M. Boyle. 2015. "Extreme work/normal work: Intensification, storytelling and hypermediation in the (re)construction of 'the New Normal.'" *Organization* 22(4): 443–456.

Grekousis, G., P. Manetos, and Y. N. Photis. 2013. "Modeling urban evolution using neural networks, fuzzy logic and GIS: The case of the Athens metropolitan area." *Cities* 30: 193–203.

Grosz, E. 2004. *The nick of time: Politics, evolution, and the untimely*. Durham, NC: Duke University Press.

Grosz, E. 2008. *Chaos, territory, art: Deleuze and the framing of the earth*. New York: Columbia University Press.

Grosz, E. 2013. "Habit today: Ravaisson, Bergson, Deleuze and us." *Body and Society* 19: 217–239.

Grover, S. 2009. "Lived experiences: Marriage, notions of love, and kinship support amongst poor women in Delhi." *Contributions to Indian Sociology* 43(1): 1–33.

Guattari, F. 1984. *Molecular revolution: Psychiatry and politics.* New York: Penguin.

Guattari, F. 1995. *Chaosmosis: An ethico-aesthetic paradigm.* Indianapolis: Indiana University Press.

Guattari, F. 2005. *The three ecologies.* London: Continuum.

Gudavarthy, A. 2016. "Brahmanism, liberalism, and the postcolonial theory." *Economic and Political Weekly* 2(24): 15–17.

Guha, R. 1983. *Elementary aspects of peasant insurgency in colonial India.* New Delhi: Oxford University Press.

Gupta, A. 2012. *Postcolonial developments: Agriculture in the making of modern India.* Durham, NC: Duke University Press.

Hakken, D. 2000. "Resocialising work? The future of the labour process." *Anthropology of Work Review* 21: 8–10.

Halpern, O. 2015. *Beautiful data: A history of vision and reason since 1945.* Durham, NC: Duke University Press.

Hanlon, G. 2012. "The 'Google model' of production—the entrepreneurial function, the immaterial and the return to rent." Paper presented at the CPPE, University of Leicester.

Hanlon, G. 2015. *The dark side of management: A secret history of management theory.* London: Routledge.

Hansen, M. B. N. 2004. *New philosophy for new media.* Cambridge, MA: MIT Press.

Haraway, D. 1987. "A manifesto for cyborgs: Science, technology, and socialist feminism in the 1980s." *Australian Feminist Studies* 2(4): 1–42.

Haraway, D. 1991. *Simians, cyborgs, and women: The reinvention of nature.* New York: Routledge.

Hardt, M. 1995. *Gilles Deleuze: An apprenticeship in philosophy.* Minneapolis: University of Minnesota Press.

Hardt, M. 1999. "Affective labor." *boundary 2* 26: 88–100.

Hardt, M., and A. Negri. 1994. *Labor of Dionysus: A critique of state-form.* Minneapolis: University of Minnesota Press.

Hardt, M., and A. Negri. 1999. "Value and affect." *boundary 2* 26(2): 77–88.

Hardt, M., and A. Negri. 2001. *Empire.* Cambridge, MA: Harvard University Press.

Hardt, M., and A. Negri. 2004. *Multitude: War and democracy in the age of empire.* New York: Penguin.

Hardt, M., and A. Negri. 2009. *Common wealth.* Cambridge, MA: Harvard University Press.

Harney, S., and F. Moten. 2013. *The undercommons: Fugitive planning and black study*. Wivenhoe: Minor Compositions.

Harrison, P. 2000. "Making sense: Embodiment and the sensibilities of the everyday." *Environment and Planning D: Society and Space* 18: 497–517.

Hart, G. 2006. "Denaturalizing dispossession: Critical ethnography in the age of resurgent imperialism." Antipode 38(5): 977–1004.

Hart, G. 2008. "The provocations of neoliberalism: Contesting the nation and liberation after apartheid." Antipode 40(4): 678–705.

Hart, K. 2007. "Phenomenality and Christianity." *Angelaki: Journal of the Theoretical Humanities* 12(1): 37–53.

Harvey, D. 2002. "The art of rent: Globalisation, monopoly and the commodification of culture." *Socialist Register* 38: 93–110.

Harvey, D. 2005. *A brief history of neoliberalism*. Oxford: Oxford University Press.

Harvey, D. 2012. *Rebel cities: From the right to the city to the urban revolution*. New York: Verso.

Hayles, N. K. 1999. *How we became posthuman: Vrtual bodies in cybernetics, literature, and informatics*. Chicago: University of Chicago Press.

Hayles, N. K. 2005. "Computing the human." *Theory, Culture and Society* 22(1): 131–151. doi: 10.1177/0263276405048438.

Heeks, R. B. 1996. *India's software industry: State policy, liberalisation and industrial development*. New Delhi: Sage.

Heidegger, M. 1962. *Being and time*. Oxford: Blackwell.

Heidegger, M. 1977. *The question concerning technology and other essays*. New York: Harper Torchbooks.

Hepworth, K. 2014. "Enacting logistical geographies." *Environment and Planning D: Society and Space* 32(6): 1120–1134.

Hesmondhalgh, D., and S. Baker. 2010. "'A very complicated version of freedom': Conditions and experiences of creative labour in three cultural industries." *Poetics* 38(1): 4–20.

Heynen, N., M. Kaika, and E. Swyngedouw, eds. 2006. *In the nature of cities: Urban political ecology and the nature of urban metabolisms*. New York: Routledge.

Hjorth, L. 2008. "Being real in the mobile reel: A case study on convergent mobile media as domesticated new media in Seoul, South Korea." *Convergence: The International Journal of Research into New Media Technologies* 14(1): 91–104.

Hochschild, A. 1983. *The managed heart: Commercialization of human feeling*. Berkeley: University of California Press.

Hoelscher, K. 2016. "The evolution of the smart cities agenda in India." *International Area Studies Review* 19: 28–44.

Holifield, R. 2009. "Actor-network theory as a critical approach to environmental justice: A case against synthesis with urban political ecology." *Antipode* 41(4): 637–658.

IMAI. 2010. *Internet and Mobile Association of India Report on Mobile Vas in India: 2010*. New Delhi: IMAI.

Jaffrelot, C. 2016. "Narendra Modi between Hindutva and subnationalism: The Gujarati asmita of a Hindu Hriday Samrat." *India Review* 15: 196–217.

Jaikumar, P. 2017. *Cinema at the end of empire: A politics of transition in Britain and India*. Durham, NC: Duke University Press.

Jamil, G. 2017. *Accumulation by segregation: Muslim localities in Delhi*. New Delhi: Oxford University Press.

Jeffrey, G. 2010. *Timepass: Youth, class, and the politics of waiting in India*. Stanford, CA: Stanford University Press.

Jiménez, A. C. 2014. "The right to infrastructure: A prototype for open source urbanism." *Environment and Planning D: Society and Space* 32(2): 342–362.

Jordan, T. 2016. "A genealogy of hacking." *Convergence: The International Journal of Research into New Media Technologies* (April): 1–17.

"Jugaad." 2018. *Oxford Living Dictionaries*, Oxford University Press. Accessed May 3, 2018. https://en.oxforddictionaries.com/definition/jugaad.

"Jugaad." 2018. *Wikipedia*. Wikimedia Foundation Inc. Accessed May 3, 2018. https://en.wikipedia.org/wiki/Jugaad.

Kahn, B., ed. 2005. *The PDMA handbook of new product development*. 2nd ed. London: John Wiley & Sons.

Kaika, M. 2005. *City of flows*. New York: Routledge.

Kaur, R. 2016. "The innovative Indian: Common man and the politics of jugaad culture." *Contemporary South Asia* 24: 313–327.

Kaviraj, S. 2010. *The imaginary institution of India: Politics and ideas*. New York: Columbia University Press.

Khan, A., and F. Ullah. 2018. "Digital populism in urban India: Exploring the motivations and modalities of urban free Wi-Fi projects in India." In *Diginaka: Digital practice in postcapitalist India*, edited by A. Rai, A. Monteiro, and K. Jayasankar. New Delhi: Orient Blackswan.

Knights, D. 2015. "Binaries need to shatter for bodies to matter: Do disembodied masculinities undermine organizational ethics?" *Organization* 22(2): 200–216.

Kracauer, S. 1995. *The mass ornament: Weimar essays*. Translated by T. Levin. Cambridge, MA: Harvard University Press.

Kraftl, P. 2014. "Liveability and urban architectures: Mol(ecul)ar biopower and the becoming lively of sustainable communities." *Environment and Planning D: Society and Space* 32(2): 274–292.

Kristeva, J. 1982. *Powers of horror: An essay on abjection*. Translated by Leon S. Roudiez. New York: Columbia University Press.

Kubitschko, S. 2015. "Hackers' media practices: Demonstrating and articulating expertise as interlocking arrangements." *Convergence: The International Journal of Research into New Media Technologies* 21(3): 388–402.

Kumar, H., and S. Bhaduri. 2014. "Jugaad to grassroot innovations: Understanding the landscape of the informal sector innovations in India." *African Journal of Science, Technology, Innovation and Development* 6(1): 13–22.

Kumar, R. 2013. "Not electricity, this is a battery revolution." *Qasba*. Accessed September 15, 2014. http://naisadak.blogspot.in/2013/12/blog-post.html ?m=1.

Kumar, R. A., dir. 2009. *Jugaad: The Movie*. Mumbai.

Kuruvilla, S., and A. Ranganathan. 2008. "Economic development strategies and macro-and micro-level human resource policies: The case of India's 'outsourcing' industry." *ILR Review* 62(1): 39–72.

Lamarca, M. G. 2014. "Federici and De Angelis on the political ecology of the commons." *Entitleblog.org*. Accessed July 10, 2016. https://entitleblog.org /2014/08/10/federici-and-de-angelis-on-the-political-ecology-of-the -commons/.

Lapoujade, D. 2010. *Powers of time*. Helsinki: n-1.

Larkin, B. 2008. *Signal and noise: Infrastructure and urban culture in Nigeria*. Durham, NC: Duke University Press.

Laslett, B., and J. Brenner. 1989. "Gender and social reproduction: Historical perspectives." *Review of Sociology* 15(1): 381–404.

Last, A. 2012. "Experimental geographies." *Geography Compass* 6(12): 706–724.

Latour, B. 2005. *Reassembling the social: An introduction to actor-network-theory*. Oxford: Oxford University Press.

Lazzarato, M. 2006. "Immaterial labor." In *Radical thought in Italy: A potential politics*, edited by M. Hardt and P. Virno, 133–150. Minneapolis: University of Minnesota Press.

Lefebvre, H. 1991. *The production of space*. Oxford: Blackwell.

Lefebvre, H. 1992. *Critique of everyday life*. Vol. 1: *Introduction*. London: Verso.

Lefebvre, H. 2002. *Critique of everyday life*. Vol. 2: *Foundations for a sociology of the everyday*. London: Verso.

Lefebvre, H. 2003. *The urban revolution*. Minneapolis: University of Minnesota Press.

Lefebvre, H. 2004. *Rhythmanalysis: Space, time and everyday life*. London: Continuum.

Lefebvre, H. 2005. *Critique of everyday life*. Vol. 3: *From modernity to modernism*. London: Verso.

Levien, M. 2012. "The land question: Special economic zones and the political economy of dispossession in India." *Journal of Peasant Studies* 39: 933–969.

Levinson, S. C. 1995. "Interactional biases in human thinking." In *Social intelligence and interaction*, edited by E. N. Goody, 221–260. Cambridge: Cambridge University Press.

Lévi-Strauss, C. 1962. *The savage mind*. Chicago: University of Chicago Press.

Leyshon, A. 2003. "Scary monsters? Software formats, peer-to-peer networks, and the spectre of the gift." *Environment and Planning D: Society and Space* 21(5): 533–558.

Liang, L. 2009. "Piracy, creativity, infrastructure: Rethinking access to culture." *Social Science Research Network*. Accessed July 14, 2016. http://papers.ssrn.com/sol3/papers.cfm?abstract_id=1436229.

Linck, M. 2008. "Deleuze's difference." *International Journal of Philosophical Studies* 16(4): 509–532. doi: 10.1080/09672550802335879.

Lindström, M. 2008. *Buy-ology: How everything we believe about why we buy is wrong*. New York: Doubleday.

Linstead, S., and T. Thanem. 2007. "Multiplicity, virtuality and organization: The contribution of Gilles Deleuze." *Organization Studies* 28(10): 1483–1501.

Lloyd-Evans, S. 2008. "Geographies of the contemporary informal sector in the global south: Gender, employment relationships and social protection." *Geography Compass* 2(6): 1885–1906.

Lodhia, S. 2015. "From 'living corpse' to India's daughter: Exploring the social, political and legal landscape of the 2012 Delhi gang rape." *Women's Studies International Forum* 50: 89–101.

Loftus, A. 2012. *Everyday environmentalism: Creating an urban political ecology*. Minneapolis: University of Minnesota Press.

Lorber, J., and L. J. Moore. 2002. *Gender and the social construction of illness*. Lanham, MD: Rowman & Littlefield.

Lorey, I. 2010. "Becoming common: Precarization as political constituting." Translated by Aileen Derieg. *e-flux journal* 17. Accessed November 19, 2016. eflux.com.

Lovink, G., and N. Rossiter. 2007. *A critique of creative industries: MyCreativity reader*. Amsterdam: Institute of Network Cultures.

Lury, C., L. Parisi, and T. Terranova. 2012. "Introduction: The becoming topological of culture." *Theory, Culture and Society* 29(4–5): 3–35.

Luxton, M., and K. Bezanson. 2006. *Social reproduction: Feminist political economy challenges neo-liberalism*. Montreal: McGill-Queen's University Press.

Lyotard, J.-F. 2014. "Energumen capitalism." In *The accelerationist reader*, edited by R. Mackay and A. Avanessian, 163–208. Falmouth, UK: Urbanomic Media.

MacKenzie, A. 2001. "The technicity of time." *Time and Society* 10(2): 235–258.

MacKinnon, D. 2011. "Reconstructing scale: Towards a new scalar politics." *Progress in Human Geography* 35(1): 21–36.

Malabou, C. 2005. *The future of Hegel: Plasticity, temporality, and dialectic*. New York: Routledge.

Malabou, C. 2008. *What should we do with our brain?* New York: Fordham University Press.

Mandal, B. C. 2013. "Dalit feminist perspectives in India." *Voice of Dalit* 6(2): 123–135.

Mandarini, M. 2005. "Antagonism, contradiction, time: Conflict and organization in Antonio Negri." *Sociological Review* 53: 192–214.

Mankekar, P. 2015. *Unsettling India: Affect, temporality, transnationality.* Durham, NC: Duke University Press.

Manning, E. 2009. *Relationscapes: Movement, art, philosophy.* Cambridge, MA: MIT Press.

Manning, E. 2013. *Always more than one: Individuation's dance.* Durham, NC: Duke University Press.

Manning, E., and B. Massumi. 2014. *Thought in the act: Passages in the ecology of experience.* Durham, NC: Duke University Press.

Manovich, L. 2001. *The language of new media.* Cambridge, MA: MIT Press.

Martin, R. 2013. "After economy? Social logics of the derivative." *Social Text* 31(1): 83–106.

Marx, K. 1861–1863. "Chapter XV: Ricardo's theory of surplus-value: The connection between Ricardo's conception of surplus-value and his views on profit and rent." https://www.marxists.org/archive/marx/works/1863/theories-surplus-value/ch15.htm.

Marx, K. 1973 [1939]. *Grundrisse: Foundations of the critique of political economy.* New York: Penguin.

Marx, K. 1976 [1867]. *Capital: A critique of political economy.* Vol. 1. New York: Penguin.

Marx, K. 1992. *Capital: A critique of political economy.* Vol. 2. London: Penguin.

Marx, K. 2010. "The Eighteenth Brumaire of Louis Bonaparte." www.marxists.org.

Mashelkar, R. 2014. "The culture of jugaad: For and against." www.beyondjugaad.com.

Massey, D. 2004. "Geographies of responsibility." *Geografiska Annaler, Series B: Human Geography* 86(1): 5–18.

Massey, D. 2005. *For space.* London: Sage.

Massumi, B. 1993. *The politics of everyday fear.* Minneapolis: University of Minnesota Press.

Massumi, B. 2002. *Parables for the virtual: Movement, affect, sensation.* Durham, NC: Duke University Press.

Massumi, B. 2011. *Semblance and event: Activist philosophy and the occurrent arts.* Boston: MIT Press.

Massumi, B. 2015a. *Ontopower: War, powers, and the state of perception.* Durham, NC: Duke University Press.

Massumi, B. 2015b. *The power at the end of the economy.* Durham, NC: Duke University Press.

Matanhelia, P. 2010. "Mobile phone use by young adults in India: A case study." PhD diss. University of Maryland, College Park.

Mayur. 2009, March 1. "'Airtel fair usage policy' sets a limit on unlimited plans." Accessed July 22, 2018. http://webtrickz.com.

Mbembe, A. 2001. *On the postcolony.* Berkeley: University of California Press.

McCarthy, J. 2005. "Devolution in the woods: Community forestry as hybrid neoliberalism." *Environment and Planning A* 37: 995–1014.

McCormack, D. P. 2005. "Diagramming practice and performance." *Environment and Planning D* 23: 119–147.

McCormack, D. P. 2007. "Molecular affects in human geographies." *Environment and Planning A* 39: 359–377.

McFarlane, C. 2008. "Sanitation in Mumbai's informal settlements: State, 'slum,' and infrastructure." *Environment and Planning A* 40: 88–107.

McIntyre, M., and H. J. Nast. 2011. "Bio (necro)polis: Marx, surplus populations, and the spatial dialectics of reproduction and 'race.'" *Antipode* 43(5): 1465–1488.

McLean, H. 2014. "Digging into the creative city: A feminist critique." *Antipode* 46(3): 669–690.

McLuhan, M. 1994. *Understanding media: The extensions of man.* Cambridge, MA: MIT Press.

Menon, N. 2011. *Seeing like a feminist.* London: Penguin.

Menon, N., and A. Nigam. 2007. *Power and contestation: India since 1989.* London: Zed Books.

Mermin, L., dir. (2006) *Office Tigers.*

Merrifield, A. 2013. *Politics of the encounter: Urban theory and protest under planetary urbanization.* Athens: University of Georgia Press.

Mezzadra, S. 2008. "Taking care: Migration and the political economy of affective labor." University of Goldsmith Papers. http://www.goldsmiths.ac.uk/csisp/papers/mezzadra_taking_care.pdf.

Mezzadra, S. 2011. "How many histories of labour? Towards a theory of postcolonial capitalism." *Postcolonial Studies* 14(2): 151–170.

Michels, C., and C. Steyaert. 2017. "By accident and by design: Composing affective atmospheres in an urban art intervention." *Organization* 24(1): 79–104.

Mitchell, T. 2008. "Rethinking economy." *Geoforum* 39(3): 1116–1121.

Moore, J. W. 2015. *Capitalism in the web of life: Ecology and the accumulation of capital.* London: Verso Books.

Morrissey, J. 2011. "Closing the neoliberal gap: Risk and regulation in the long war of securitization." *Antipode* 43(3): 874–900.

Moten, F. 2003. *In the break: The aesthetics of the black radical tradition.* Minneapolis: University of Minnesota Press.

Munster, A. 2006. *Materializing new media: Embodiment in information.* Hanover, NH: Dartmouth University Press.

Münster, D., and C. Strümpell. 2014. "Anthropology of neoliberal India." *Contributions to Indian Sociology* 48: 1–16.

Mutch, A. 2016. "The limits of process: On (re)reading Henri Bergson." *Organization* 23(6): 825–839.

Nagendra, H. 2016. *Nature in the city: Bengaluru in the past, present, and future.* New Delhi: Oxford University Press.

Nandy, A. 1998. *The secret politics of our desires: Innocence, culpability and Indian popular cinema.* London: Palgrave Macmillan.

Narrain, A. 2008. "That despicable specimen of humanity": Policing of homosexuality in India." In *Challenging the rule(s) of law: Colonialism, criminology and human rights in India*, edited by K. Kannabiran and R. Singh, 48–77. New Delhi: Sage.

Nayar, P. 2012. *Digital cool: Life in the age of new media.* Hyderabad: Orient Blackswan.

Negri, A. 1999. "Value and affect." Translated by M. Hardt. *boundary 2* 26(2): 77–88.

Negri, A. 2004. *Subversive Spinoza.* Edited by T. S. Murphy. Manchester: Manchester University Press.

Neveling, P. 2014. "Structural contingencies and untimely coincidences in the making of neoliberal India: The Kandla Free Trade Zone, 1965–91." *Contributions to Indian Sociology* 48: 17–43.

Nietzsche, F. 1966. *Beyond good and evil.* Translated by W. Kaufmann. New York: Vintage.

Nietzsche, F. 1983. *Untimely meditations.* Cambridge: Cambridge University Press.

Nietzsche, F. 1999. *The birth of tragedy and other writings.* Cambridge: Cambridge University Press.

Niranjana, T. 1992. *Siting translation: History, post-structuralism, and the colonial context.* Berkeley: University of California Press.

Nolan, P., and J. Zhang. 2007. *The global business revolution and the cascade effect: Systems integration in the global aerospace, beverage and retail industries.* Basingstoke, UK: Palgrave Macmillan.

Odendaal, N. 2006. "Towards the digital city in South Africa: Issues and constraints." *Journal of Urban Technology* 13: 29–48.

Paltasingh, T., and L. Lingam. 2014. "'Production' and 'reproduction' in feminism: Ideas, perspectives and concepts." *IIM Kozhikode Society and Management Review* 3(1): 45–53.

Pandey, G. 1990. *The construction of communalism in colonial North India.* New Delhi: Oxford University Press.

Parayil, G. ed. 2016. *Political economy and information capitalism in India: Digital divide, development divide and equity.* Springer.

Parikka, J. 2012. *What is media archaeology?* Cambridge, UK: Polity.

Parisi, L. 2004. *Abstract sex: Philosophy, biotechnology, and mutations of desire.* London: Continuum.

Parisi, L. 2013. *Contagious architecture computation, aesthetic and space.* Cambridge, MA: MIT Press.

Parisi, L., and T. Terranova. 2000. "Heat-death: Emergence and control in genetic engineering and artificial life." *CTheory* 84. Accessed November 19, 2016. http://www.ctheory.netltexcfile.asp ?pick= 127.

Parthasarathy, B. 2005. "The political economy of the computer software industry in Bangalore, India." In *ICTs and Indian economic development: Economy, work, regulation,* edited by M. A. A. V. Saith, 199–230. New Delhi: Sage.

Patel, G. 2017. *Risky bodies and techno-intimacy: Reflections on sexuality, media, science, finance.* Seattle: University of Washington Press.

Peirce, C. S. 1998. *The essential Peirce: Selected philosophical writings, 1893–1913.* Bloomington: Indiana University Press.

Pelbart, P. P. 2011. "The deterritorialised unconscious." In *The Guattari effect,* edited by E. Alliez and A. Coffrey, 68–83. New York: Continuum.

Peticca-Harris, A., J. Weststar, and S. McKenna. 2015. "The perils of project-based work: Attempting resistance to extreme work practices in video game development." *Organization* 22(4): 570–587.

Phadke, S., S. Khan, and S. Ranade. 2009. "Why loiter? Radical possibilities for gendered dissent." In *Dissent and cultural resistance: In Asia's cities,* edited by M. Butcher and S. Velayutham. London: Routledge.

Phadke, S., S. Khan, and S. Ranade. 2011. *Why loiter? Women and risk on Mumbai streets.* New Delhi: Penguin.

Pierce, J. 1995. *Gender trials: Emotional lives in contemporary law firms.* Berkeley: University of California Press.

Pile, S. 1990. "Depth hermeneutics and critical human geography." *Environment and Planning D* 8: 211–232.

Poster, M., and D. Savat. 2009. *Deleuze and new technology.* Edinburgh: Edinburgh University Press.

Posteraro, T. S. 2014. "Organismic spatiality: Toward a metaphysic of composition." *Environment and Planning D: Society and Space* 32(4): 739–752.

Pousttchi, K., and Y. Hufenbach. 2011. "Value creation in the mobile market." *Business and Information Systems Engineering* 5: 299–312.

Povinelli, E. 2016. *Geontologies: A requiem to late liberalism.* Durham, NC: Duke University Press.

Prakash, G. 1990. *Bonded histories: Genealogies of labour servitude in colonial India.* Cambridge: Cambridge University Press.

Prasad, A. 2009. "Capitalizing disease biopolitics of drug trials in India." *Theory, Culture and Society* 26: 1–29.

Prasad, S. 2015. "Sanitizing the domestic: Hygiene and gender in late colonial Bengal." *Journal of Women's History* 27(3): 132–153.

Prashad, V. 2002. *Everybody was kung fu fighting: Afro-Asian connections and the myth of cultural purity*. Boston: Beacon.

Pratt, A. C. 2009. "Urban regeneration: From the arts 'feel good' factor to the cultural economy. A case study of Hoxton, London." *Urban Studies* 46(5–6): 1041–1061.

Pratt, G., C. Johnston, and V. Banta. 2017. "Lifetimes of disposability and surplus entrepreneurs in Bagong Barrio, Manila." *Antipode* 49(1): 169–192.

Precarious Workers Brigade. 2015. "Training for exploitation." *Journal of Aesthetics and Protest Press*. Available at http://joaap.org/press/pwb/PWB _TrainingForExploitation_smaller.pdf.

Puar, J. 2009. *Terrorist assemblages: Homonationalism in queer times*. Durham, NC: Duke University Press.

Puar, J. 2017. *The right to maim: Debility, capacity, disability*. Durham, NC: Duke University Press.

Pullen, A., and C. Rhodes. 2015. "Ethics, embodiment and organizations." *Organization* 22(2): 159–165.

Pullen, A., C. Rhodes, and T. Thanem. 2017. "Affective politics in gendered organizations: Affirmative notes on becoming-woman." *Organization* 24(1): 105–123.

Radjou, N., J. Prabhu, and S. Ahuja. 2012. *Jugaad innovation: Think frugal, be flexible, generate breakthrough growth*. San Francisco: Jossey-Bass.

Rai, A. S. 1997. "'Thus spake the subaltern . . . ': Postcolonial criticism and the scene of desire." *Discourse: Journal for Theoretical Studies in Media and Culture* 19(2): 163–183.

Rai, A. S. 2009. *Untimely Bollywood: Globalization and India's new media assemblage*. Durham, NC: Duke University Press.

Rangaswamy, N., and S. Nair. 2010. "The mobile phone store ecology in a Mumbai slum community: Hybrid networks for enterprise." *Information Technologies and International Development* 16(3): 51–65.

Rangaswamy, N., and N. Sambasivan. 2011. "Cutting Chai, jugaad, and hera pheri: Towards UbiComp for a global community." *Perspective on Ubiquitous Computing* 15(6): 553–564.

Ravaisson, F. 2008. *Of habit*. Translated by C. Carlisle and M. Sinclaire. London: Bloomsbury.

Rawat, R. S. 2015. "Genealogies of the Dalit political: The transformation of Achhut from 'Untouched' to 'Untouchable' in early twentieth-century North India." *Indian Economic and Social History Review* 52(3): 335–355.

Rawat, R. S., and K. Satyanarayana. 2016. *Dalit studies*. Durham, NC: Duke University Press.

Robinson, C. 2000. *Black Marxism: The making of the black radical tradition*. Chapel Hill: University of North Carolina Press.

Roelvink, G., and M. Zolkos. 2015. "Affective ontologies: Post-humanist perspectives on the self, feeling and intersubjectivity." *Emotion, Space and Society* 14: 47–49.

Rossi, U. 2015. "The variegated economics and the potential politics of the smart city." *Territory, Politics, Governance* 4: 337–353.

Rossiter, N., J. Tretschok, C. Wiedemann, and S. Zehle, eds. 2012. *Depletion desire: A glossary of network ecologies.* Amsterdam: Institute of Network Cultures.

Routledge, P. 2008. "Acting in the network: ANT and the politics of generating associations." *Environment and Planning D: Society and Space* 26(2): 199–217.

Roy, A. 2001. "A 'public' muse: On planning convictions and feminist contentions." *Journal of Planning Education and Research* 21(2): 109–126.

Roy, A. 2005. "Urban informality: Toward an epistemology of planning." *Journal of the American Planning Association* 71(2): 147–158.

Roy, A. 2009. "Why India cannot plan its cities: Informality, insurgence and the idiom of urbanization." *Planning Theory* 8(1): 76–87.

Roychowdhury, P. 2013. "'The Delhi gang rape': The making of international causes." *Feminist Studies* 39(1): 282–292.

Sabat, H. 2008. "Why different carriers adopt different spectrum acquisition strategies." *Information Technology and Management* 9: 251–284.

Safri, M. 2015. "Mapping noncapitalist supply chains: Toward an alternate conception of value creation and distribution." *Organization* 22(6): 924–941.

Safri, M., and J. Graham. 2010. "The global household: Toward a feminist postcapitalist international political economy." *Signs* 36(1): 99–126.

Sage, D., P. Fussey, and A. Dainty. 2015. "Securing and scaling resilient futures: Neoliberalization, infrastructure, and topologies of power." *Environment and Planning D: Society and Space* 33(3): 494–511.

Saha, D., and J. Sen. 2016. "Understanding clustering in creative-knowledge cities: Creative clusters in Kolkata, India." *GSTF Journal of Engineering Technology (JET)* 3: 33–38.

Sahlins, M. 1972. *Stone age economics.* New York: Aldine de Gruyter.

Saldanha, A. 2007. *Psychedelic white: Goa trance and the viscosity of race.* Minneapolis: University of Minnesota Press.

Sanyal, K. 2007. *Rethinking capitalist development: Primitive accumulation, governmentality, and post-colonial capitalism.* London: Routledge.

Sanyal, K., and R. Bhattacharyya. 2009. "Beyond the factory: Globalisation, informalisation of production, and the new locations of labour." *Economic and Political Weekly* 44: 35–44.

Sanyal, K., and R. Bhattacharyya. 2011. "Bypassing the squalor: New towns, immaterial labour and exclusion in postcolonial urbanisation." *Economic and Political Weekly* 46: 41–49.

Sartre, J.-P. 2004. *Critique of dialectical reason.* Vol. 1: *Theory of practical ensembles.* Translated by A. Sheridan-Smith. London: Verso.

Savat, D. 2009. "Deleuze and technology." In *Deleuze and technology*, edited by M. Poster and D. Savat, 1–15. Edinburgh: Edinbugh University Press.

Schumpeter, J. A. 2008. *The theory of economic development: An inquiry into profits, capital, credit, interest, and the business cycle.* New Brunswick, NJ: Transaction.

Scott, A. 2011. "A world in emergence: Notes towards a resynthesis of urban economic geography." *Urban Geography* 32(6): 845–870.

Scott, J. C. 1985. *Weapons of the weak: Everyday forms of peasant resistance.* New Haven, CT: Yale University Press.

Sen, S., and B. Dasgupta. 2009. *Unfreedom and waged work: Labour in India's manufacturing industry.* New Delhi: Sage.

Sen, S., and P. Nair. 2004. "A report on trafficking in women and children in India 2002–2003." *Methodology* 33: 39.

Sharma, U. 1986. *Women's work, class and the urban household: A study of Shimla, North India.* London: Tavistock.

Sharpe, J. 1993. *Allegories of empire: The figure of woman in the colonial text.* Minneapolis: University of Minnesota Press.

Shaviro, S. 2014. *The universe of things: On speculative realism.* Minneapolis: University of Minnesota Press.

Simpson, P. 2008. "Chronic everyday life: Rhythmanalysing street performance." *Social and Cultural Geography* 9: 807–829.

Simpson, P. 2009. "'Falling on deaf ears': A postphenomenology of sonorous presence." *Environment and Planning A* 41: 2556–2575.

Simpson, P. 2013. "Ecologies of experience: Materiality, sociality, and the embodied experience of (street) performing." *Environment and Planning A* 45(1): 180–196.

Singh, C. P., A. Lather, and D. P. Goyal. 2009. "Building relationships @ BPO India." *Paradigm (Institute of Management Technology)* 13: 110–114.

Singh, J. A. 2011. "Cinema and the underdog: On an underground video parlour and the life filmic." *Caravanmagazine.in.* Accessed December 15, 2017. http://www.caravanmagazine.in/reviews-and-essays/cinema-and-underdog #sthash.1ryKOlUb.dpuf.

Singh, N. 2016. "Let's start a revolution . . . but how?" *Why Loiter?* (blog). Accessed December 2, 2016. http://whyloiter.blogspot.com/.

Sinha, M. 2000. "Refashioning Mother India: Feminism and nationalism in late-colonial India." *Feminist Studies* 26(3): 623–644.

Slavova, M., and E. Okwechime. 2016. "African smart cities strategies for agenda 2063." *Africa Journal of Management* 2: 210–229.

Smiers, J. 2007. "What if we would not have copyright? New business models for cultural entrepreneurs." In *MyCreativity reader: A critique of creative industries*, edited by N. Rossiter and G. Lovink, 191–206. Amsterdam: Institute of Network Cultures.

Smith, S. L. 1995. *Sick and tired of being sick and tired: Black women's health activism in America, 1890–1950.* Philadelphia: University of Pennsylvania Press.

Soat, M. 2018. "Just a little nudge." *Marketing News* 52(1): 3. Business Source Complete, EBSCOhost.

Sohn-Rethel, A. 1978. *Intellectual and manual labour: A critique of epistemology.* London: Macmillan; Atlantic Highlands, NJ: Humanities Press.

Sommers, T., L. Shields, and J. MacLean. 1987. "Older women's league (U.S.)." In *Task force on caregivers: Women take care: The consequences of caregiving in today's society.* Gainesville, FL: Triad.

Spinoza, B. 1992. *Ethics.* Translated by S. Shirley. Edited by S. Feldman. Cambridge, MA: Hackett.

Spivak, G. C. 1999. *A critique of postcolonial reason.* Cambridge, MA: Harvard University Press.

Spivak, G. C. 2012. *In other worlds: Essays in cultural politics.* London: Routledge.

Srinivasan, T. N. 2005. "Information technology and India's growth prospects." In *Offshoring white-collar work—the issues and implications,* edited by L. A. C. Brainard. Washington, DC: Brookings Institution.

Srivastava, S. 2007. *Passionate modernity: Sexuality, class, and consumption in India.* New Delhi: Routledge.

Staples, D. E. 2006. *No place like home: Organizing home-based labor in the era of structural adjustment.* New York: Routledge.

Starosta, G. 2012. "Cognitive commodities and the value-form." *Science and Society* 76(3): 365–392.

Stiegler, B. 1998. *Technics and time: The fault of Epimetheus.* Stanford, CA: Stanford University Press.

Streeck, W. 2014. *Buying time: The delayed crisis of democratic capitalism.* New York: Verso Books.

Strumińska-Kutra, M. 2016. "Engaged scholarship: Steering between the risks of paternalism, opportunism, and paralysis." *Organization* 23(6): 864–883.

Sundaram, R. 2009. *Pirate modernity: Delhi's media urbanism.* London: Routledge.

Swyngedouw, E. 2006. "Metabolic urbanization: The making of cyborg cities." In *In the Nature of Cities,* edited by N. Heynen, M. Kaika, and E. Swyngedouw, 20–39. London: Routledge.

Tagore, R. 2002. *Gora.* New Delhi: Rupa Publications.

Tambe, A. 2000. "Colluding patriarchies: The colonial reform of sexual relations in India." *Feminist Studies* 26(3): 586–600.

Tarde, G. 2012. *Monadology and sociology.* Translated by Theo Lorenc. New York: re.press.

Terranova, T. 2004. *Network culture: Politics for the information age.* New York: Pluto Press.

Thaler, R. H., and C. R. Sunstein. 1999. *Nudge: Improving decisions about health, wealth, and happiness.* New Haven, CT: Yale University Press.

Thanem, T., and L. Wallenberg. 2015. "What can bodies do? Reading Spinoza for an affective ethics of organizational life." *Organization* 22(2): 235–250.

Thoburn, N. 2008. "What is a militant?" In *Deleuze and Politics*, edited by I. Buchanan and N. Thoburn, 98–120. Edinburgh: Edinburgh University Press.

Thrift, N. 2004. "Intensities of feeling: Towards a spatial politics of affect." *Geografiska Annaler, Series B* 86: 57–78.

Thrift, N. 2005. "From born to made: Technology, biology and space." *Transactions of the Institute of British Geographers* 30(4): 463–476.

Thrift, N. 2006. "Re-inventing invention: New tendencies in capitalist commodification." *Economy and Society* 35: 279–306.

Thrift, N. 2007. *Non-representational theory: Spaces, politics, affects.* London: Routledge.

Thrift, N. 2014. "The 'sentient'city and what it may portend." *Big Data and Society* 1(1): 1–21.

Titus, N. T. Forthcoming. "The other cinemas: Recycled content, vulnerable bodies, and the gradual dismantling of publicness." In *Diginaka: Where the digital meets the local in India*, edited by Anjali Monteiro, K. P. Jayasankar, and Amit Rai. New Delhi: Orient Blackswan.

Toscano, A. 2008. "The open secret of real abstraction." *Rethinking Marxism* 20(2): 273–287. doi: 10.1080/08935690801917304.

Tronti, M. 2005. "The strategy of refusal—Mario Tronti." Accessed May 25, 2016. https://libcom.org/library/strategy-refusal-mario-tronti.

Udupa, T. 2017. "When jugaad is justified." *The Hindu*, May 22, 2017. http://www.thehindu.com/sci-tech/when-startups-should-resort-to-jugaad-and-when-its-not-worth-it/article18524922.ece.

Uno, K. 2012. *The genesis of an unknown body.* Translated by Melissa McMahon. Helsinki: n-1.

Upadhya, C. 2008. "Review of *Virtual migration: The programming of globalization.*" *Contributions to Indian Sociology* 42: 344–347.

Upadhya, C. 2009. "Controlling offshore knowledge workers: Power and agency in India's software outsourcing industry." *New Technology, Work and Employment* 24: 2–18.

Valayden, D. 2016. "Racial feralization: Targeting race in the age of 'planetary urbanization.'" *Theory, Culture and Society* 33(7–8): 159–182.

Vanita, R. ed. 2013. *Queering India: Same-sex love and eroticism in Indian culture and society.* New York: Routledge.

Varela, F. J. 1999. *Ethical know-how: Action, wisdom, and cognition.* Stanford, CA: Stanford University Press.

Varela, F. J., E. Thompson, and E. Rosch. 1991. *The embodied mind: Cognitive science and human experience.* Cambridge, MA: MIT Press.

Vasudevan, R. S., J. Bagchi, R. Sundaram, M. Narula, G. Lovink, and S. Sengupta, eds. 2002. *Sarai reader 02: The cities of everyday life*. New Delhi: Sarai/SONM.

Venkatraman, S. 2017. *Social media in South India*. London: UCL Press.

Vercellone, C. 2007. "From formal subsumption to general intellect: Elements for a Marxist reading of the thesis of cognitive capitalism." *Historical Materialism* 15(1): 13–36.

Vindhya, U., and V. Dev. 2011. "Survivors of sex trafficking in Andhra Pradesh: Evidence and testimony." *Indian Journal of Gender Studies* 18(2): 129–165.

Virilio, P. 2005. *The information bomb*. London: Verso.

Virno, P. 2003. *A grammar of the multitude: For an analysis of contemporary forms of life*. Los Angeles: Semiotext(e).

Waitt, G., E. Ryan, and C. Farbotko. 2013. "A visceral politics of sound." *Antipode* 46(1): 283–300.

Wang, D., S. Park, and D. R. Fesenmaier. 2012. "The role of smartphones in mediating the touristic experience." *Journal of Travel Research* 51(4): 371–387.

Warf, B. 1990. "The reconstruction of social ecology and neighborhood change in Brooklyn." *Environment and Planning D* 8: 73–96.

Warf, B. 2007. "Oligopolization of global media and telecommunications and its implications for democracy." *Ethics, Place and Environment* 10(1): 89–105.

Wark, M. 2012. *Telesthesia: Communication, culture, and class*. Cambridge, UK: Polity.

Webb, J. 2004. "Organizations, self-identities and the new economy." *Sociology* 38(4): 719–738.

Webb, M. 2013. "Disciplining the everyday state and society? Anti-corruption and right to information activism in Delhi." *Contributions to Indian Sociology* 47(3): 363–393.

Weeks, K. 2011. *The problem with work: Feminism, Marxism, antiwork politics, and postwork imaginaries*. Durham, NC: Duke University Press.

Westlund, O. 2010. "New(s) functions for the mobile: A cross-cultural study." *New Media and Society* 12(1): 91–108.

Whitehead, A. N. 1979. *Process and reality: An essay in cosmology*. New York: Free Press.

Williams, R. 1977. *Marxism and literature*. London: Oxford University Press.

Williams, S., S. Katz, and P. Martin. 2011. "The neuro-complex: Some comments and convergences." *Media Tropes* 3(1): 135–146.

Williamson, B. 2016. "Computing brains: Learning algorithms and neurocomputation in the smart city." *Information, Communication and Society* 20: 81–99.

Wilson, D. 2005. "Gentrification, discourse, and the body: Chicago's Humboldt Park." *Environment and Planning D: Society and Space* 23: 295–331.

Wood, W., and D. Rünger. 2016. "Psychology of habit." *Annual Review of Psychology* 67(1): 289–314.

Wyly, E. 2012. "The city of cognitive-cultural capitalism." *City* 17(3): 387–394.

Yang, G. Z., L. Dempere-Marco, X. P. Hu, and A. Rowe. 2002. "Visual search: Psychophysical models and practical applications." *Image and Vision Computing* 20(4): 291–305. doi: 10.1016/s0262-8856(02)00022-7.

Yusoff, K. 2018. "Politics of the Anthropocene: Formation of the commons as a geologic process." *Antipode* 50(1): 255–276.

Zeiderman, A. 2016. "Submergence: Precarious politics in Colombia's future port-city." *Antipode* 48(3): 809–831.

Zurawicki, L. 2010. *Neuromarketing: Exploring the brain of the consumer.* New York: Springer.

Index

actual, x–xiii, xiv, 2, 7, 18–19, 31–32, 40, 171n1; as counteractualization, xv–xvi, 33–35, 43, 60. *See also* virtual

aesthetic, 70, 162, 168n4

affect: abduction and, 4; in affective ethnography, 3, 8, 17, 31, 50, 64–66, 72, 77, 79, 155; consumption and, 13, 29; digital media and, 26, 50, 66; duration and, 18, 30, 50–51, 62, 77; ecology and, 22, 24, 30, 65–66, 81–82, 107, 121; as embodied, 17, 21, 51, 72; emotion and, 18; as environment, 23, 78, 128–130, 135, 138, 147; essence and, 18–19, 50, 119; habit and, 100; jugaad and, xvi–xvii, 31, 112; labor and, 72, 81–82, 88, 97–98, 107–108, 111–114, 126; labor of women and, 123; method and, ix, 2, 7, 10, 14, 19, 22, 31, 46, 49, 60, 79, 119, 142; neoliberalism and, 22, 29, 46–47, 59–60, 74, 90; as nonhuman, 16, 18, 50; pathetic image and, 26; politics and, 1, 22, 32–33, 35–37, 70, 81–82, 127, 137, 152, 156, 159, 164; representation and, 18, 46, 83; as rhythm analyses, 76, 108–110, 144; in smart cities, 147; technicity and, 46, 53, 65, 91. *See also* actual; affective ethnography; assemblage; attention; becoming; biopolitical production; cognitive labor; Deleuze, G.; diagrammatics; ecology of sensation; Guattari, F.; habit/habituation; jugaad; mobile phones; neoliberalism; virtual

affective ethnography, x–xi; 1, 3, 6, 8, 17, 19, 20, 31–33, 50, 64–66, 69, 72–73, 77–79, 155. *See also* affect

affirmation, 1, 11, 22, 32–33, 35–37, 127, 137, 152, 156, 159, 164

agency, 7, 11–12, 16, 19, 122, 135, 142

Ahmed, S., 17

Anthropocene, 27, 171n2

Ash, J., 53, 59–60, 65–66

assemblage: digital media and, 4, 69, 72, 79, 82–83, 88–90, 122, 129; jugaad and, 2, 91, 112; method and, 72, 162; as techno-perceptual, 101, 107, 130

attention: decolonizing and, 16–17, 35, 50; ecology of sensation and, xvi, 3, 6, 16–17, 35, 50–51, 60, 69–70, 100, 122–123, 140–142, 147, 160, 165; jugaad and, 50. *See also* affect; affective ethnography; assemblage; biopolitical production; cognitive labor; Deleuze, G.; diagrammatics; ecology of sensation; Guattari, F.; habit/habituation; jugaad; mobile phones; neoliberalism

Barbagallo, C., 27

becoming, ix–x, 5, 8, 11, 15, 31, 32, 38–40, 61, 64, 70, 77, 82–83, 87, 96, 101, 105–107, 120–122, 127–130, 140, 151, 155. *See also* actual; affect; assemblage; biopolitical production; Deleuze, G.; diagrammatics; ecology of sensation; Guattari, F.; habit/habituation; jugaad; revolutionary becoming; virtual

Bharatiya Janata Party (BJP), 2, 29. *See also* Hindutva

Bharti Airtel, 52–57, 61–62

Big Data, 23, 69, 83, 101, 138–140, 146

biopolitical production, 3, 16, 24–25, 31–32, 48, 71, 76, 107–109, 110, 112, 119, 126, 128–129, 135, 142, 144, 146, 147–148, 172n1. *See also* affect; affective ethnography; assemblage; biopolitical production; cognitive labor; Deleuze, G.; ecology of sensation; Guattari, F.; habit/habituation; jugaad; mobile phones; neoliberalism; ontology; plasticity; value

business process outsourcing (BPO), 83–86; and masculinism, 86

capitalism, xi, xiii, 1, 2, 6–8, 10, 14, 15, 46, 52, 67, 71–73, 77–78, 81–83, 87, 98–100, 106–107, 113, 115, 126–127, 129, 137–139, 146–147, 149, 151, 154, 157. *See also* actual; affect; affective ethnography; assemblage; biopolitical production; cognitive labor; ecology of sensation; neoliberalism; value

City and Industrial Development Corporation of Maharashtra LTD (CIDCO), 73, 171n4

Clough, P., xvii, 33, 69–70, 128

cognitive labor: knowledge economy and, 24, 76, 87; manual labor and, 7, 20, 22, 86–87, 94, 97; neoliberalism and, 1, 3, 7–8, 11, 25, 82–85, 90, 112, 122. *See also* affect; affective ethnography;

assemblage; biopolitical production; Deleuze, G.; ecology of sensation; Guattari, F.; habit/habituation; jugaad; mobile phones; neoliberalism; ontology; plasticity; value

colonialism, ix, xv, 17, 26, 52, 58, 66, 69, 76, 112, 121, 126, 145

commons: as commoning, 28, 65, 82, 119, 123–126, 155; as common notion, 160; and mobile phones, 143; and violence, 28;

creativity: capture of, 136; jugaad and, 4, 10, 12, 19, 33, 48, 57, 71, 88

Dalits: and caste, 79, 116; and domestic labor, 142; and gender, 145; and media ecologies, 24–26, 74–75, 110, 142; and smart cities, 9, 37, 74–75, 111;

Deleuze, G., x–xi, 6, 16–18, 26–27, 72, 106, 154; and critique of capital, 22, 34; and revolutionary becoming, 157–158

diagram. *See* diagrammatics

diagrammatics: and empiricism, 3, 6, 18, 22, 24, 28, 48, 58, 62, 71–72, 106, 135, 162; and Guattari, 9–10, 169n.13; and jugaad, 32, 49, 63, 155; and language, 155; as method, xi–xii, xvi, 9–10, 23, 36–37, 60–63, 69–70, 82, 126, 130, 135, 142, 156; and parataxis, 7, 73, 83. *See also* affect; affective ethnography; assemblage; biopolitical production; cognitive labor; Deleuze, G.; ecology of sensation; Guattari, F.; habit/ habituation; jugaad; mobile phones; neoliberalism; ontology; plasticity; value

digital media: and control, 11, 64–65, 82–83, 87–88, 98–99, 141, 147–148; as ecology of sensation, 21, 64, 77–79; in India, 77, 84–86; and neoliberalism, 2, 50, 80, 167n3; and social reproduction, 115 ecology of sensation, x, xv, 3, 6, 18, 22, 24, 28, 48, 58, 62–63, 71–72; affect and, 9, 11, 31, 46, 50–52, 58, 60, 88–90, 126, 140; digital media and, 6, 61, 77, 108–109; as environment, 23, 78, 128–130, 135, 138, 147; jugaad and, 4, 16, 32, 34, 46, 57, 65, 91, 161; neoliberalism and, 77, 99–100; smart cities and, 83. *See also* affect; affective ethnography; assemblage; biopolitical production; cognitive labor; Deleuze, G.; diagrammatics; Guattari, F.; habit/habituation; jugaad; mobile phones; neoliberalism

emancipation, xii, 12, 21, 149, 155, 160–161, 172n2

entrepreneurialism, xiii, 6, 20–22, 25–27, 30, 32, 46, 48, 70–71, 76, 81, 111, 126, 137, 144, 147, 156, 159, 171n2. *See also* cognitive labor; ecology of sensation; habit/habituation; jugaad; mobile phones; neoliberalism; value

ethics, x–xi, 10, 14, 28, 34, 37

financialization, 135, 160, 172n4

Florida, R., 95, 118, 129, 136, 140

Foucault, M., xiii, 13, 47, 71, 110, 123, 140, 146

gender: critique of affect and, 22, 65, 68–69, 71–100, 146; and media practice, 2–3, 68, 70; unpaid labor of women and, 82, 113–114, 127. *See also* affect; biopolitical production; neoliberalism; ontology; value

Ghar/Bahir (home/world), 108, 112, 114, 121–123, 127

Guattari, F., xi, 16–17, 36–37, 154, 168n13

habit/habituation, x, xvi, 3, 6, 16–17, 35, 50–51, 58, 60, 69–70, 72, 81, 90, 100, 122–123, 127, 128, 140, 143, 147, 160,

165, 168n8; and becoming, 8; and smart cities, 137, 141–142, 148–149; and time, 143, 153. *See also* affect; biopolitical production; cognitive labor; Deleuze, G.; ecology of sensation; Guattari, F.; jugaad; mobile phones; neoliberalism; ontology; plasticity

hacking: as ecology, 6, 27–29, 64, 111, 115–117, 132, 142, 149; and empiricism x, 7; and habituation, 6, 63, 69–70; and jugaad, 1, 4, 6, 32, 34, 46, 65, 90–91, 161, 164–165; and media ecologies, 5, 91, 165; neoliberalism and, 21, 32, 118, 154. *See also* affect; affective ethnography; assemblage; biopolitical production; cognitive labor; Deleuze, G.; ecology of sensation; Guattari, F.; habit/habituation; jugaad; mobile phones; neoliberalism; ontology; plasticity; value

Harney, S., 27, 154, 171n6

Harvey, D., 21, 27

Hindu chauvinism. *See* Hindutva

Hindutva, 6, 37, 46, 119, 137, 145, 170n1

immaterial labor. *See* cognitive labor

informal economy: jugaad and, 3, 8, 22, 27, 32, 65, 83, 142; mobile phones and, 120; "organized" sector and, 7, 121; piracy and, 7; smart cities and, 139. *See also* biopolitical production; cognitive labor; ecology of sensation; hacking; jugaad; mobile phones; neoliberalism; piracy; value

innovation, 87, 106

jugaad: as abduction, 4–5; affect and, 26, 46; Bharti Airtel and, 55–57; and becoming, 155–156; control and, 125; creativity and, 4, 10, 12, 19, 33, 48, 57, 71, 88; definition of, x, xiii, 30–31, 51, 54–55, 91, 161, 168n8, 170n1;

jugaad (*continued*)
   diagram of, 32, 49, 63, 155; ecology and, 3, 9, 11, 31, 46, 52, 58, 88–90, 126, 140; as essentially Indian, 5, 15, 19; as ethos, x, 168n6; as event, 19–20, 45, 49–51, 55, 58–59, 65, 158; as extralegal, 46, 58, 91–92; as frugal innovation, 32, 48; habit and, 16–17, 35, 50–51, 58, 60, 69–70, 72, 81, 90, 100, 122–123, 127, 128, 140, 143, 147, 160, 165; as know-how, 5, 8; labor and, 81; language and, 51, 54, 64, 154; as life-hacking, 107; method and, 4, 16, 32, 34, 46, 65, 91, 161; mobile phones and, 73–74, 90, 132; misogyny and, 2, 28, 163, 165, 168n10; neoliberalism and, 2, 6, 8, 11, 14, 21–22, 32, 48, 55–56, 65–67, 118; ontology and, 30, 114, 160; patriarchy and, 11, 22, 165; patronage and, 9; politics and, 32–33, 156; as practice, ix, 2, 7, 10, 14, 22, 32, 45–47, 49, 61, 119, 142; process and, 93; queerness and, 32–33, 35–37, 70, 81–82, 127, 152, 156, 159, 164; smart cities and, 23, 88; as statecraft, 116; time and, 5, 51, 55, 63, 92–94, 117, 153, 156; virtual and, 32; virtuosity and, 27, 125. *See also* affect; affective ethnography; assemblage; biopolitical production; cognitive labor; Deleuze, G.; ecology of sensation; Guattari, F.; habit/habituation; jugaad; mobile phones; neoliberalism; ontology; plasticity; value
Jugaad Innovation, 44, 48

logistics, 6, 7, 27–29, 64–65, 76, 87, 110, 111–113, 115, 119, 131–132, 139, 144–145, 148–149, 154. *See also* affective ethnography; assemblage; biopolitical production; ecology of sensation; jugaad; mobile phones; neoliberalism; value

manual labor. *See* cognitive labor
Massumi, B., 36, 143, 152, 167n2

media ecologies, 3–4, 7–8, 11, 27–29, 34, 52, 58, 63, 79, 89, 91–92. *See also* hacking; mobile phones; smart cities
metabolic imbroglios, 111, 115, 131–133, 147. *See also* biopolitical production; cognitive labor; ecology of sensation; jugaad; mobile phones; neoliberalism; ontology; plasticity; value
mobile phones: and biopolitical production, 3–4, 119; and consumption, 4, 15, 55–57, 60–62, 79, 89, 91–92, 142, 172n6; and control, 4, 96; and ecologies of sensation, 3, 108–109, 133, 136, 170n1; gender and, 2–3, 68, 70; and habituation, 3–5, 52, 69, 75–76, 89, 119, 133, 143; and Hindi cinema, 87; and jugaad, 73–74, 89–90, 117, 132, 145; and labor, 88; and memory, 173n5; and misogyny, 2–3; and mobile value-added services (MVAS), 3–4, 53–55, 132; and neoliberalism, 3–4, 52, 60, 73–74, 79, 128; and photography, 75–76; and repair wallahs, 32, 132–134; as smartphones, 3–4, 15, 52, 73–75, 87–90, 135–136, 138–139. *See also* affect; affective ethnography; assemblage; biopolitical production; cognitive labor; ecology of sensation; habit/habituation; jugaad; neoliberalism; value
Moten, F., 27, 157

Negri, A., 80–81, 172n3
neoliberalism: biopolitical production and, 3, 16, 24–25, 31–32, 48, 71, 76, 107–109, 110, 112, 119, 126, 128–129, 135, 142, 144, 146, 147–148, 172n1; business process outsourcing (BPO) and, 85–88; and control, 112; as depoliticizing, 1, 137; and digital media, 1, 3, 7–8, 11, 25, 82–85, 90, 112, 122; division of labor and, 98, 108, 114; entrepreneurialism and, 14, 37, 48, 25–27, 137; India

and, 52, 69, 136–137; jugaad and, 21, 32, 118, 154; liberation and, 130; monopoly and, 21, 29; precarity and, 158; process and, 3, 92; security and, 3, 97; smart cities and, 135–136, 138–139; space and, 96; time and, 3, 92–94; value and, 140, 142. *See also* affect; biopolitical production; cognitive labor; ecology of sensation; habit/habituation; jugaad; mobile phones; plasticity; value

Nietzsche, F., 34

nonlinear dynamics, 34–35, 51–52, 63, 72–73, 79–80, 85, 100

Office Tigers, 84–87

ontology: affect and, 6, 9–10, 14–16, 18–20, 23, 27, 29, 34, 36–38, 40, 49, 114; hacking and, 7, 32, 49, 63, 120, 125, 129–130, 135, 140, 148–149, 155; method and, 34, 37, 100, 106, 110, 121–123, 126–127, 154. *See also* affect; affective ethnography; assemblage; biopolitical production; cognitive labor; Deleuze, G.; ecology of sensation; Guattari, F.; habit/habituation; hacking; jugaad; mobile phones; neoliberalism; nonlinear dynamics; plasticity; value

"organized" sector. *See* informal economy

parataxis, x, 3, 7, 23, 31, 72, 88, 168n4; and intensity, 10

patriarchy, 2, 11, 22, 28, 145, 163–165. *See also* gender

piracy, 1, 45, 57, 69, 111, 122. *See also* biopolitical production; cognitive labor; ecology of sensation; habit/habituation; informal economy; jugaad; mobile phones; neoliberalism; plasticity; value

plasticity, 6, 27–29, 30, 64, 78, 111, 114–117, 132–134, 140–142, 147–149, 155, 160. *See also* affect; affective ethnography; assemblage; biopolitical production; cognitive labor; Deleuze, G.; ecology of sensation; Guattari, F.; habit/habituation; jugaad; mobile phones; neoliberalism; ontology; smart cities; value

postcolonial studies, 1, 46, 59, 82; and the smart city, 129; and the subaltern, 109

Povinelli, E., 15–16, 31, 47, 66, 78, 167n2

precarity, 128, 143, 159

Puar, J., xvii, 71, 128, 156

representational space. *See* smart cities: and representational space

revolutionary becoming, 8–9, 140–142, 147–149, 155–156. *See also* becoming

Simondon, G., 6, 167

smart cities: and affective environments, 23, 78, 128–130, 135, 138, 147; and Big Data, 23, 138; definition of, 23, 138; and diagrammatics, 135, 142; and digital media, 83, 90; as elite project, 81–83; and entrepreneurialism, 126, 131–133; and Narendra Modi, 36; and neoliberalism, 76, 138; plasticity of, 78, 134, 140–142, 147; and representational space, 119; and undercommons, 73, 88. *See also* biopolitical production; cognitive labor; ecology of sensation; jugaad; mobile phones; neoliberalism; ontology; plasticity; value

social reproduction, 11–13, 106, 112–113; digital media and, 90, 114, 121; domestic labor and, 116–117; as domestic mode of production, 172n2; hacking and, 115; smart cities and, 129. *See also* biopolitical production; cognitive labor; hacking; jugaad; mobile phones; neoliberalism; ontology; value

Sohn-Rethel, A., 7, 20, 106

www.ingramcontent.com/pod-product-compliance
Lightning Source LLC
Chambersburg PA
CBHW070323270326
41926CB00017B/3733